Linux开发书系

Linux
云计算
Kubernetes实战

吴光科 程 浩 刑亚飞 编著

清华大学出版社
北京

内 容 简 介

本书从实用的角度出发，详细介绍了 Kubernetes 的相关理论与应用，包括 Kubernetes 组件概念、Kubernetes 云计算平台配置实战、Kubernetes 企业网络 Flannel 实战、Kubernetes 核心组件 services 实战、Kubernetes Pod 容器升级实战、Kubernetes+NFS 持久化存储实战、Kubernetes+CephFS 持久化存储实战、Kubernetes+Ceph RBD 持久化存储实战、Prometheus 监控 Kubernetes 实战、Kubernetes etcd 服务实战、Kubernetes+HAProxy 高可用集群和 Kubernetes 配置故障实战。

本书免费提供与书中内容相关的视频课程讲解，以指导读者深入地进行学习，详见前言中的说明。

本书既可作为高等学校计算机相关专业的教材，也可作为系统管理员、网络管理员、Linux 运维工程师及网站开发、测试、设计等人员的参考用书。

本书封面贴有清华大学出版社防伪标签，无标签者不得销售。
版权所有，侵权必究。举报：010-62782989，beiqinquan@tup.tsinghua.edu.cn。

图书在版编目（CIP）数据

Linux云计算：Kubernetes实战 / 吴光科，程浩，刑亚飞编著. —北京：清华大学出版社，2023.5
（Linux开发书系）
ISBN 978-7-302-63429-4

Ⅰ．①L… Ⅱ．①吴… ②程… ③刑… Ⅲ．①Linux操作系统—程序设计 Ⅳ．①TP316.85

中国国家版本馆CIP数据核字（2023）第080587号

责任编辑：	刘　星
封面设计：	李召霞
责任校对：	李建庄
责任印制：	丛怀宇

出版发行：清华大学出版社
　　　　网　　　址：http://www.tup.com.cn, http://www.wqbook.com
　　　　地　　　址：北京清华大学学研大厦A座　　邮　　编：100084
　　　　社　总　机：010-83470000　　邮　　购：010-62786544
　　　　投稿与读者服务：010-62776969, c-service@tup.tsinghua.edu.cn
　　　　质　量　反　馈：010-62772015, zhiliang@tup.tsinghua.edu.cn
　　　　课　件　下　载：http://www.tup.com.cn, 010-83470236
印　装　者：北京同文印刷有限责任公司
经　　　销：全国新华书店
开　　　本：186mm×240mm　　印　　张：11.25　　字　　数：215千字
版　　　次：2023年7月第1版　　印　　次：2023年7月第1次印刷
印　　　数：1～2000
定　　　价：69.00元

产品编号：101570-01

前言
PREFACE

Linux 是当今三大操作系统（Windows、macOS、Linux）之一，其创始人是林纳斯·托瓦兹[①]。林纳斯·托瓦兹 21 岁时用 4 个月的时间首次创建了 Linux 内核，于 1991 年 10 月 5 日正式对外发布。Linux 系统继承了 UNIX 系统以网络为核心的思想，是一个性能稳定的多用户网络操作系统。

20 世纪 90 年代至今，互联网飞速发展，IT 引领时代潮流，而 Linux 系统是一切 IT 的基石，其应用场景涉及方方面面，小到个人电脑、智能手环、智能手表、智能手机等设备，大到服务器、云计算、大数据、人工智能、数字货币、区块链等领域。

为什么写《Linux 云计算——Kubernetes 实战》这本书？这要从我的经历说起。我出生在贵州省一个贫困的小山村，从小经历了砍柴、放牛、挑水、做饭、日出而作、日落而归的朴素生活，看到父母一辈子都生活在小山村里，没有见过大城市，所以从小立志要走出大山，要让父母过上幸福的生活。正是这样的信念让我不断地努力。大学毕业至今，我在"北漂"的 IT 运维路上已走过了十多年：从初创小公司到国有企业、机关单位，再到图吧、研修网、京东商城等 IT 企业，分别担任过 Linux 运维工程师、Linux 运维架构师、运维经理，直到现在创办的京峰教育培训机构。

一路走来，很感谢生命中遇到的每一个人，是大家的帮助，让我不断地进步和成长，也让我明白了一个人活着不应该只为自己和自己的家人，还要考虑到整个社会，哪怕只能为社会贡献一点点价值，人生就是精彩的。

为了帮助更多的人通过技术改变自己的命运，我决定和团队同事一起编写这本书。虽然市面上关于 Linux 的书籍有很多，但是很难找到一本关于 Kubernetes 组件概念、Kubernetes 平台配置实战、Kubernetes 企业网络 Flannel 实战、Kubernetes 核心组件 services 实战、Kubernetes Pod 容器升级实战、Kubernetes+NFS 持久化存储实战、Kubernetes+CephFS 持久化存储实战、

[①] 创始人全称是 Linus Benedict Torvalds（林纳斯·本纳第克特·托瓦兹）。

Kubernetes+Ceph RBD 持久化存储实战、Prometheus 监控 Kubernetes 实战、Kubernetes etcd 服务实战、Kubernetes+HAProxy 高可用集群和 Kubernetes 配置故障实战等的详细、全面的主流技术书籍，这就是编写本书的初衷。

配套资源

- 程序代码、面试题目、学习路径、工具手册、简历模板等资料，请扫描下方二维码下载或者到清华大学出版社官方网站本书页面下载。

配套资源

- 作者精心录制了与 Linux 开发相关的视频课程（3000 分钟，144 集），便于读者自学。扫描封底"文泉课堂"刮刮卡中的二维码进行绑定后即可观看（注：视频内容仅供学习参考，与书中内容并非一一对应）。

虽然已花费大量的时间和精力核对书中的代码和内容，但难免存在纰漏，恳请读者批评指正。

吴光科

2023 年 4 月

致 谢
ACKNOWLEDGEMENT

感谢 Linux 之父 Linus Benedict Torvalds，他不仅创造了 Linux 系统，还影响了整个开源世界，也影响了我的一生。

感谢我亲爱的父母，含辛茹苦地抚养我们兄弟三人，是他们对我无微不至的照顾，让我有更多的精力和动力去工作，去帮助更多的人。

感谢吴俊、李芬伦、陈权志、胡智超、焦伟、曾地长、孙峰、黄超、赵敬星、曾令军、张杰、丁刘倩、刘到波、班风仑及其他挚友多年来对我的信任和鼓励。

感谢腾讯课堂所有的课程经理及平台老师，感谢 51CTO 副总裁一休及全体工作人员对我及京峰教育培训机构的大力支持。

感谢京峰教育培训机构的每位学员对我的支持和鼓励，希望他们都学有所成，最终成为社会的中流砥柱。感谢京峰教育首席运营官蔡正雄，感谢京峰教育培训机构的辛老师、朱老师、张老师、关老师、兮兮老师、小江老师、可馨老师等全体老师和助教、班长、副班长，是他们的大力支持，让京峰教育能够帮助更多的学员。

最后要感谢我的爱人黄小红，是她一直在背后默默地支持我、鼓励我，让我有更多的精力和时间去完成这本书。

吴光科
2023 年 4 月

目录 CONTENTS

第1章 Kubernetes 组件概念 ... 1
- 1.1 云计算概念 ... 1
- 1.2 云计算技术的分类 ... 1
- 1.3 Kubernetes 入门及概念介绍 ... 2
- 1.4 Kubernetes 平台组件概念 ... 3
- 1.5 Kubernetes 工作原理剖析 ... 4
- 1.6 Pod 概念剖析 ... 7
- 1.7 label 概念剖析 ... 8
- 1.8 Replication Controller 概念剖析 ... 8
- 1.9 service 概念剖析 ... 9
- 1.10 node 概念剖析 ... 10
- 1.11 Kubernetes volume 概念剖析 ... 10
- 1.12 Deployment 概念剖析 ... 11
- 1.13 DaemonSet 概念剖析 ... 11
- 1.14 StatefulSet 概念剖析 ... 11
- 1.15 ConfigMap 概念剖析 ... 12
- 1.16 Secrets 概念剖析 ... 13
- 1.17 CronJob 概念剖析 ... 14
- 1.18 Kubernetes 证书剖析和制作实战 ... 15

第2章 Kubernetes 云计算平台配置实战 ... 25
- 2.1 Kubernetes 节点 hosts 及防火墙设置 ... 25
- 2.2 Linux 内核参数设置和优化 ... 26
- 2.3 Docker 虚拟化案例实战 ... 26
- 2.4 Kubernetes 添加部署源 ... 27
- 2.5 Kubernetes Kubeadm 案例实战 ... 28
- 2.6 Kubernetes master 节点实战 ... 30
- 2.7 Kubernetes 集群节点和删除 ... 31
- 2.8 Kubernetes 节点网络配置 ... 31
- 2.9 Kubernetes 开启 IPVS 模式 ... 39
- 2.10 Kubernetes 集群故障排错 ... 40

2.11 Kubernetes 集群节点移除 40
2.12 etcd 分布式案例操作 40

第 3 章 Kubernetes 企业网络 Flannel 实战 42
3.1 Flannel 工作原理 42
3.2 Flannel 架构介绍 43
3.3 Kubernetes Dashboard UI 实战 44
3.4 Kubernetes YAML 文件详解 47
3.5 kubectl 常见指令操作 49
3.6 Kubernetes 本地私有仓库实战 50

第 4 章 Kubernetes 核心组件 service 实战 52
4.1 Kubernetes service 概念 52
4.2 Kubernetes service 实现方式 53
4.3 service 实战：ClusterIP 案例演练 54
4.4 service 实战：NodePort 案例演练 55
4.5 service 实战：LoadBalancer 案例演练 56
4.6 service 实战：Ingress 案例演练 58
4.7 Kubernetes Traefik 案例实战 63

第 5 章 Kubernetes 容器升级实战 73
5.1 Kubernetes 容器升级概念 73
5.2 Kubernetes 容器升级实现方式 73
5.3 Kubernetes 容器升级测试 75
5.4 Kubernetes 容器升级验证 76
5.5 Kubernetes 容器升级回滚 77
5.6 Kubernetes 滚动升级和回滚原理 78

第 6 章 Kubernetes+NFS 持久化存储实战 82
6.1 Kubernetes 服务运行状态 82
6.2 Kubernetes 存储系统 83
6.3 Kubernetes 存储绑定的概念 84
6.4 PV 的访问模式 84
6.5 Kubernetes+NFS 静态存储模式 86
6.6 PVC 存储卷创建 87
6.7 Nginx 整合 PV 存储卷 88
6.8 Kubernetes+NFS 动态存储模式 90
6.9 NFS 插件配置实战 91

第 7 章 Kubernetes+CephFS 持久化存储实战 96
7.1 Kubernetes+CephFS 静态存储模式 96

7.2 PV 存储卷创建 ... 96
7.3 PVC 存储卷创建 ... 97
7.4 Nginx 整合 CephFS PV 存储卷 ... 98
7.5 Kubernetes+CephFS 动态存储模式 ... 100
7.6 CephFS 动态插件配置实战 ... 101

第 8 章 Kubernetes+Ceph RBD 持久化存储实战 ... 105
8.1 Kubernetes+Ceph RBD 静态存储模式 ... 105
8.2 PV 存储卷创建 ... 105
8.3 PVC 存储卷创建 ... 107
8.4 Nginx 整合 Ceph PV 存储卷 ... 107
8.5 Kubernetes+Ceph RBD 动态存储模式 ... 109
8.6 Ceph RBD 插件配置实战 ... 110

第 9 章 Prometheus 监控 Kubernetes 实战 ... 117
9.1 Prometheus 监控优点 ... 117
9.2 Prometheus 监控特点 ... 118
9.3 Prometheus 组件实战 ... 118
9.4 Prometheus 体系结构 ... 119
9.5 Prometheus 工作流程 ... 120
9.6 Prometheus 和 Kubernetes 背景 ... 120
9.7 Kubernetes 集群部署 node-exporter ... 121
9.8 Kubernetes 集群部署 Prometheus ... 122
9.9 Kubernetes 集群部署 Grafana ... 129
9.10 Kubernetes 配置和整合 Prometheus ... 131
9.11 Kubernetes+Prometheus 报警设置 ... 135
9.12 Kubernetes Alertmanager 实战 ... 136
9.13 Alertmanager 实战部署 ... 140

第 10 章 Kubernetes etcd 服务实战 ... 146
10.1 etcd 和 ZK 服务概念 ... 146
10.2 etcd 的使用场景 ... 147
10.3 etcd 读写性能 ... 147
10.4 etcd 工作原理 ... 147
10.5 etcd 选主 ... 148
10.6 etcd 日志复制 ... 148
10.7 etcd 安全性 ... 149
10.8 etcd 使用案例 ... 150
10.9 etcd 接口使用 ... 150

第 11 章　Kubernetes+HAProxy 高可用集群 ……………………………………… 151

11.1　Kubernetes 高可用集群概念 ……………………………………………………… 151
11.2　Kubernetes 高可用工作原理 ……………………………………………………… 151
11.3　HAProxy 安装配置 ………………………………………………………………… 152
11.4　配置 Keepalived 服务 ……………………………………………………………… 156
11.5　Keepalived master 配置实战 ……………………………………………………… 157
11.6　Keepalived Backup 配置实战 ……………………………………………………… 158
11.7　创建 HAProxy 检查脚本 …………………………………………………………… 160
11.8　HAProxy+Keepalived 验证 ………………………………………………………… 160
11.9　初始化 master 集群 ………………………………………………………………… 161
11.10　Kubernetes Dashboard UI 实战 …………………………………………………… 163

第 12 章　Kubernetes 配置故障实战 …………………………………………………… 166

12.1　etcd 配置中心故障错误一 ………………………………………………………… 166
12.2　etcd 配置中心故障错误二 ………………………………………………………… 167
12.3　Pod infrastructure 故障错误三 …………………………………………………… 167
12.4　Docker 虚拟化故障错误四 ………………………………………………………… 168
12.5　Docker 虚拟化故障错误五 ………………………………………………………… 168
12.6　Dashboard API 故障错误六 ………………………………………………………… 168
12.7　Dashboard 网络访问故障错误七 ………………………………………………… 169

第 1 章 Kubernetes 组件概念

1.1 云计算概念

云计算技术其实是将硬件设备、操作系统、软件服务、网络带宽、流量、计费系统等资源组成一个大的资源池(动态扩容、弹性伸缩),然后可以再将所有的资源池分配给租户使用,租户可以根据自身的需求,按需购买资源。

云计算技术强调的是资源池,是租户的概念。虚拟化技术是云计算技术框架中的一个小模块、组件技术。云计算技术最终的产物包括硬件设备、操作系统、软件服务、网络带宽等。每个产物都可以租给用户使用,用户可以自行购买。

对于云计算技术的资源池,租户不需要了解云计算底层框架、架构,只要清楚自身对资源池的需求,需要多少台服务器、多少云主机、多大网络带宽,最终按需付费即可。

1.2 云计算技术的分类

云计算技术按照实现方式分为 3 种类型:基础设施即服务(Infrastructure as a Service,IaaS)、平台即服务(Platform as a Service,PaaS)、软件即服务(Software as a Service,SaaS),如图 1-1 所示。每种云的特点如下。

(1)基础设施即服务。

① 租户无须管理底层硬件设备、网络、服务器、存储、虚拟化技术。

② 租户只需对操作系统、中间件、数据、应用做维护即可。

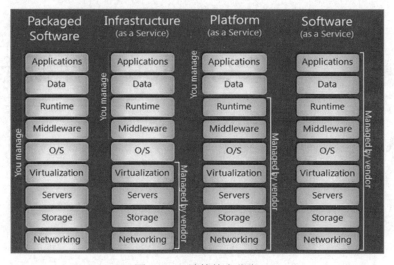

图 1-1 云计算技术分类

（2）平台即服务。

① 租户无须管理底层硬件设备、网络、服务器、存储、虚拟化技术、操作系统、中间件。

② 租户只需对应用服务、软件程序做维护，无须操作系统和底层设施。

（3）软件即服务。

① 租户无须管理底层硬件设备、网络、服务器、存储、虚拟化技术、操作系统、中间件、应用服务、软件程序等。

② 租户只需花钱、付费，提交业务需求，运营商将满足租户所有需求。

1.3 Kubernetes 入门及概念介绍

Kubernetes 又称为 K8S（首字母为 K，首字母与尾字母之间有 8 个字符。尾字母为 S，所以简称 K8S）或者简称为 kube，是一种可以自动实施 Linux 容器操作的开源平台。

Kubernetes 可以帮助用户省去应用容器化过程中的许多手动部署和扩展操作。我们可以将运行 Linux 容器的多组主机聚集在一起，由 Kubernetes 帮助用户轻松高效地管理这些集群，而且这些集群可跨公共云、私有云或混合云部署主机。因此，对于要求快速扩展的云原生应用而言（例如，借助 Apache Kafka 进行的实时数据流处理），Kubernetes 是理想的托管平台。

Kubernetes 最初由 Google 公司的工程师开发和设计。Google 是最早研发 Linux 容器技术的企业之一（组建了 cgroups），曾公开分享介绍 Google 如何将一切都运行于容器之中（这是 Google

云服务背后的技术）。Google 每周会启用超过 20 亿个容器——全都由内部平台 Borg 支撑。Borg 是 Kubernetes 的前身，多年来开发 Borg 的经验教训成了影响 Kubernetes 中许多技术的主要因素。

在企业生产环境中，应用会涉及多个容器（主机），这些容器必须跨多个服务器主机进行部署。容器安全性需要多层部署，因此可能比较复杂。但 Kubernetes 有助于解决这一问题。Kubernetes 可以提供所需的编排和管理功能，以便针对这些工作负载大规模部署容器。借助 Kubernetes 编排功能，可以构建跨多个容器的应用服务、跨集群调度、扩展这些容器，并长期持续管理这些容器的健康状况。

Kubernetes（K8S）是自动化容器操作的开源平台，这些操作包括部署、调度和节点集群间扩展。如果用户曾经用过 Docker 容器技术部署容器，可以将 Docker 看成 Kubernetes 内部使用的低级别组件。Kubernetes 不仅支持 Docker，还支持 Rocket，这是另一种容器技术。使用 Kubernetes 可以实现如下功能：

（1）自动化容器的部署和复制。
（2）跨多台主机进行容器编排和管理。
（3）有效管控应用部署和更新，并实现自动化操作。
（4）挂载和增加存储，用于运行有状态的应用。
（5）能够快速、按需扩展容器化应用及其资源。
（6）对服务进行声明式管理，保证部署的应用始终按照部署的方式运行。
（7）更加充分地利用硬件，最大程度获取运行企业应用所需的资源。
（8）利用自动布局、自动重启、自动复制及自动扩展功能，对应用实施状况进行检查和自我修复。

1.4　Kubernetes 平台组件概念

Kubernetes 集群中主要存在两种类型的节点：master 节点和 node 节点。

1）master 节点

master 节点主要负责对外提供一系列管理集群的 API 接口，并且通过与 node 节点交互实现对集群的操作管理。以下为 master 节点的组件和用途。

（1）Apiserver：用户和 Kubernetes 集群交互的入口，封装了核心对象的增、删、改、查操作，提供了 RESTFul 风格的 API 接口，通过 etcd 实现持久化并维护对象的一致性。

（2）Scheduler：负责集群资源的调度和管理。例如，当有 Pod 异常退出需要重新分配机器时，Scheduler 通过一定的调度算法找到最合适的节点。

（3）controller-manager：主要用于保证 replication controller 定义的复制数量和实际运行的 Pod（容器）数量一致，并保证从 service 到 Pod 的映射关系总是最新的。

2）node 节点

node 节点主要负责接收 master 发送的指令，同时与本地 Docker 引擎交互等操作。以下为 node 节点的组件和用途。

（1）kubelet：运行在 node 节点，负责与节点上的 Docker 交互。例如，启停容器、监控运行状态等。

（2）kube-proxy：运行在 node 节点上，负责为 Pod 提供代理功能，会定期从 etcd 获取 service 信息，并根据 service 信息通过修改 iptables 实现流量转发（最初的版本是直接通过程序提供转发功能，效率较低），将流量转发到要访问的 Pod 所在的节点上。

Kubernetes 云计算平台除了 master 和 node 节点上相应的组件模块之外，还需要 etcd、Flannel、Docker 等插件。以下为相关插件的用途。

（1）etcd：etcd 是一个分布式一致性 key-value 存储系统数据库，可用于注册与发现配置共享和服务，用来存储 Kubernetes 的信息。etcd 组件作为一个高可用、强一致性的服务发现存储仓库，渐渐为开发人员所关注。在云计算时代，如何让服务快速透明地接入计算集群，如何让共享配置信息快速被集群中的所有机器发现，更为重要的是，如何构建这样一套高可用、安全、易于部署以及响应快速的服务集群？etcd 的诞生就是为了解决该问题。

（2）Flannel：Flannel 是 CoreOS 团队针对 Kubernetes 设计的一个覆盖网络（Overlay Network）工具，目的是为集群中的所有节点重新制定 IP 地址的使用规则，从而使不同节点上的容器能够获得同属一个内网且不重复的 IP 地址，并让属于不同节点上的容器能够直接通过内网 IP 通信。

（3）Docker：Docker 是一款轻量级的虚拟化软件，主要是为了解决企业轻量级服务器操作系统和应用容器而诞生的，在 Kubernetes 集群中，Docker 被看成 Kubernetes 集群中的低级别组件，负责接收 node 节点上 kubelet 组件进行交互操作，如启动、停止、删除 Docker 容器，监控 Docker 容器运行状态等。

1.5 Kubernetes 工作原理剖析

Kubernetes 集群是一组节点，这些节点可以是物理服务器或者虚拟机，在其上安装 Kubernetes 平台。Kubernetes 云计算架构如图 1-2 所示，图中为了强调核心概念有所简化。

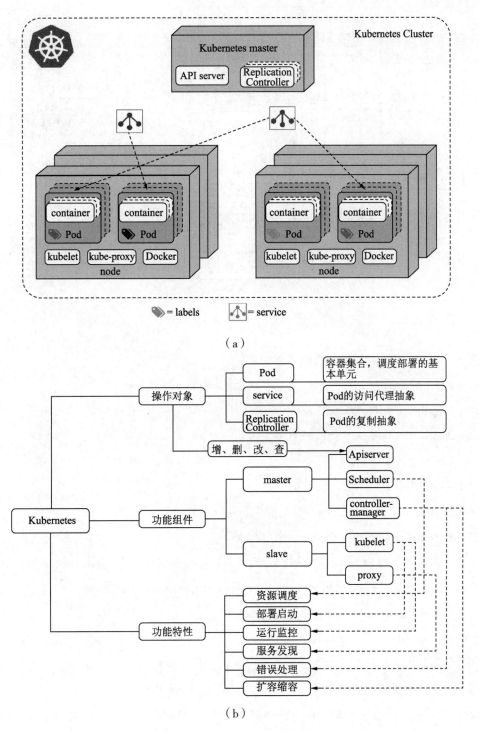

图 1-2　Kubernetes 云计算架构

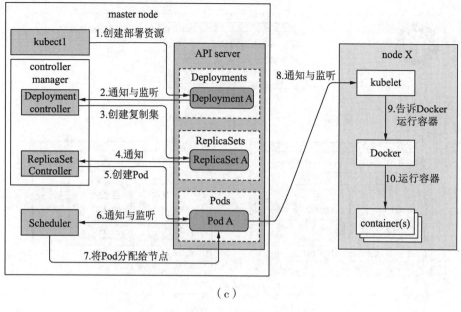

（c）

图 1-2 （续）

图 1-2（a）中相关模块和组件的名称如下：

（1）Pod（容器组）。

（2）container（容器）。

（3）labels(🏷)（标签）。

（4）service（🖥）（服务）。

（5）node（Kubernetes 计算节点）。

（6）Replication Controller（副本控制器）。

（7）Kubernetes master（Kubernetes 主节点）。

从图 1-2 可以看到 Kubernetes 组件和逻辑关系：Kubernetes 集群主要由 master 和 node 两类节点组成；master 的组件包括 API server、controller-manager、Scheduler 和 etcd 等，其中 API server 是整个集群的网关。

Kubernetes node 主要由 kubelet、kube-proxy、Docker 引擎等组成。kubelet 是 Kubernetes 集群的工作与节点上的代理组件。

在企业生产环境中，完整的 Kubernetes 集群还包括 CoreDNS、Prometheus（或 HeapSter）、Dashboard、Ingress Controller、cAdvisor 等几个附加组件。其中 cAdvisor 组件作用于各个节点（master 和 node 节点），用于收集容器和节点的 CPU、内存以及磁盘资源的利用率指标数据，这些统计数据由 Heapster 聚合后，可以通过 API server 访问。

Kubernetes 集群中创建一个资源（Pod 容器）的工作流程和步骤如下。

（1）客户端提交创建（Deployment、Namespace、Pod）请求，可以通过 API server 的 Restful API 提交，也可以使用 kubectl 命令行工具提交。

（2）通过 API server 处理用户的请求，并将相关数据（Deployment、Namespace 和 Pod）存储到 etcd 配置数据库中。

（3）Kubernetes Scheduler 调度器通过 API server 查看未绑定的 Pod，并尝试为该 Pod 分配 node 主机资源。

（4）过滤主机（调度预选）：调度器用一组规则过滤掉不符合要求的主机。例如，Pod 指定了所需要的资源量，那么可用资源比 Pod 需要的资源量少的主机就会被过滤掉。

（5）主机打分（调度优选）：对第一步筛选出的符合要求的主机进行打分。在主机打分阶段，调度器会考虑一些整体优化策略，例如，把 Replication Controller 的副本分布到不同的主机上，使用最低负载的主机等。

（6）选择主机：选择打分最高的主机，进行 binding 操作，将结果存储到 etcd 中。

（7）node 节点上的 kubelet 根据调度结果，调用主机上的 Docker 引擎执行 Pod 创建操作，绑定成功后，Scheduler 会调用 API server 的 API 在 etcd 中创建一个 boundpod 对象，描述在一个工作节点上绑定运行的所有 Pod 信息。

（8）同时运行在每个工作节点上的 kubelet 也会定期与 etcd 同步 boundpod 信息，一旦发现应该在该工作节点上运行的 boundpod 对象没有更新，将调用 Docker API 创建并启动 Pod 内的容器。

1.6 Pod 概念剖析

Pod（容器组）是 Kubernetes 中最小的可布署单元了，一般部署在 node 节点上，包含一组容器和卷。同一个 Pod 中的容器共享同一个网络命名空间，可以使用 localhost 互相通信，Pod 是短暂的，不是持续性实体。关于 Pod 的常见问题和解答如下：

（1）Pod 一般是短暂的，如何才能持久化容器数据，使其能够跨重启而存在呢？Kubernetes 支持卷的概念，因此可以使用持久化的卷类型。

（2）如果要创建同一个容器的多份副本，需要逐个创建出来吗？可以手动创建单个 Pod，也可以使用 Replication Controller 中的 Pod 模板创建出多份副本。

（3）Pod 是短暂的，重启 Pod 时 IP 地址可能会改变，那么怎样才能从前端容器正确可靠地指向后台容器呢？答案是可以使用 service。

1.7　label 概念剖析

Kubernetes label 是标记到 Pod 的一个键/值对，用来传递用户定义的属性。例如，用户可能创建了一个 tier 和 app 标签，可通过 label（tier=frontend, app=jfedu-app）标记前端 Pod 容器，使用 label（tier=backend, app=jfedu-app）标记后台 Pod。

可以使用 Selectors 选择带有特定 label(标签)的 Pod,并且将 service 或者 Replication Controller 应用于这些 Pod。

1.8　Replication Controller 概念剖析

Replication Controller 确保任意时间都有指定数量的 Pod "副本"在运行。如果为某个 Pod 创建了 Replication Controller 并指定 3 个副本，它会创建 3 个 Pod，并持续监控它们。如果某个 Pod 不响应，那么 Replication Controller 会替换它，并保持总数为 3，如图 1-3 所示。

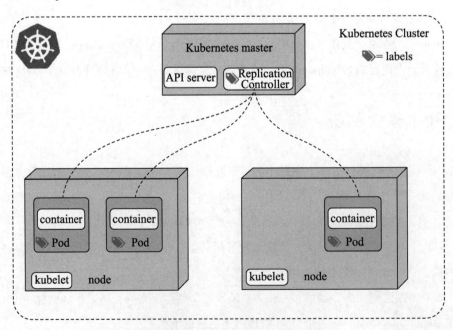

图 1-3　Replication Controller 副本结构

如果之前不响应的 Pod 恢复，就有了 4 个 Pod，那么 Replication Controller 会将其中一个 Pod 终止，以保持总数为 3。如果在运行中将副本总数改为 5，Replication Controller 会立刻启动 2 个新 Pod，保证总数为 5。还可以按照这样的方式减少 Pod 副本，这个特性在执行滚动升级时很有用。当创建 Replication Controller 时，需要指定以下两个内容。

（1）Pod 模板：用来创建 Pod 副本的模板。

（2）label：Replication Controller 需要监控的 Pod 的标签。

1.9　service 概念剖析

service 是定义一系列 Pod 以及访问这些 Pod 的策略的一层抽象。service 通过 label 找到 Pod 组。因为 service 是抽象的，所以通常看不到它们的存在，这就让这一概念更难以理解。Kubernetes service 内部结构如图 1-4 所示。

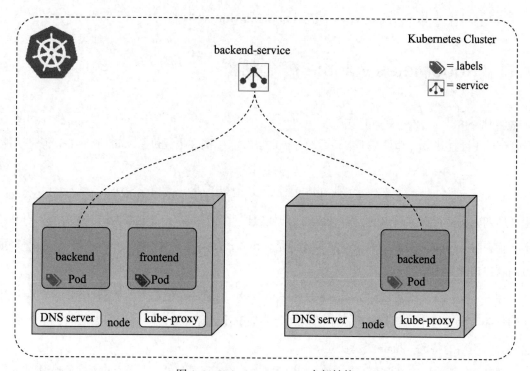

图 1-4　Kubernetes service 内部结构

假设创建了 2 个 Pod 容器，并定义后台 service 的名称为 backend-service，label 选择器为

（tier=backend, app=jfedu-app），backend-service 的 service 会完成如下操作：

（1）会为 service 创建一个本地集群的 DNS 入口，因此前端 Pod 只需要 DNS 查找主机名为 backend-service，就能够解析出前端应用程序可用的 IP 地址。

（2）service 为创建的 2 个 Pod 容器提供透明的负载均衡，并将请求分发给其中的任意一个，通过每个 node 上运行的代理（kube-proxy）完成。

1.10　node 概念剖析

node（节点）是物理或虚拟机器，作为 Kubernetes worker，通常称为计算节点。每个节点都运行以下 Kubernetes 关键组件。

（1）kubelet：是 node 节点上的主程序，负责接收 master 发送的指令。

（2）kube-proxy：service 使用 kube-proxy 将请求转发至 Pod。

（3）Docker 或 Rocket：Kubernetes 使用 Docker 或 Rocket 容器技术创建容器。

1.11　Kubernetes volume 概念剖析

在 Docker 中有 volume 这个概念，volume 只是磁盘上或其他容器中的简单目录，生命周期不受管理，并且直到最近都是基于本地后端存储的。Docker 现在也提供了 volume driver，但是功能较弱。

Kubernetes 的 volume 有着明显的生命周期——和使用它的 Pod 生命周期一致。因此，volume 生命周期就比运行在 Pod 中的容器要长久，即使容器重启，volume 上的数据依然保存着。当 Pod 不再存在时，volume 也就消失了。更重要的是，Kubernetes 支持多种类型的 volume，且 Pod 可以同时使用多种类型的 volume。

在 Kubernetes 内部实现中，volume 只是一个目录，目录中可能有一些数据，Pod 的容器可以访问这些数据。这个目录是如何产生的？它后端基于什么存储介质？其中的数据内容是什么？这些都由使用的特定 volume 类型决定。

要使用 volume，Pod 需要指定 volume 的类型和内容（spec.volumes 字段），以及映射到容器的位置（spec.containers.volumeMounts 字段）。

容器中的进程可以看到 Docker image 和 volumes 组成的文件系统。Docker image 处于文件系统架构的 root，任何 volume 都映射在镜像的特定路径上。volume 不能映射到其他 volume 上，也不能硬链接到其他 volume。容器中的每个容器必须指定它们要映射的 volume。

Kubernetes 支持多种类型的 volume，包括 emptyDir、hostPath、gcePersistentDisk、awsElasticBlockStore、nfs、iscsi、flocker、glusterfs、rbd、cephfs、gitRepo、secret、persistentVolumeClaim 等。

1.12 Deployment 概念剖析

Deployment 为 Pod 和 ReplicaSet 提供声明式更新和部署，只需要在 Deployment 中描述想要的目标状态，Deployment controller 就会将 Pod 和 ReplicaSet 的实际状态改变到目标状态。

可以定义一个全新的 Deployment 创建 ReplicaSet，或者删除已有的 Deployment 并创建一个新的替换。

注意：无须手动管理由 Deployment 创建的 ReplicaSet，否则就越俎代庖了。

1.13 DaemonSet 概念剖析

DaemonSet 对象能确保其创建的 Pod 在集群中的每一台（或指定）node 上都运行一个副本。如果集群中动态加入了新的 node，DaemonSet 中的 Pod 也会被添加到新加入的 node 上。删除一个 DaemonSet 也会级联删除所有其创建的 Pod。DaemonSet 和 Deployment 的区别如下。

（1）Deployment 部署 Pod 会分布在各个 node 上，node 可能运行好几个副本。

（2）DaemonSet 部署 Pod 会分布在各个 node 上，每个 node 只能运行一个 Pod。

以下为 DaemonSet 的使用场景：

（1）在每台节点上运行一个集群存储服务，如运行 glusterd、ceph。

（2）在每台节点上运行一个日志收集服务，如 fluentd、logstash。

（3）在每台节点上运行一个节点监控服务，如 Prometheus node exporter、collectd、Datadog agent、New Relic agent 或 Ganglia gmond。

1.14 StatefulSet 概念剖析

Kubernetes RC、Deployment 和 DaemonSet 都是面向无状态的服务，它们所管理的 Pod 的 IP、

名字、启停顺序等都是随机的，而 StatefulSet 是什么？顾名思义，StatefulSet 是有状态的集合，管理所有有状态的服务，如 MySQL、MongoDB 集群等。

StatefulSet 本质上是 Deployment 的一种变体，在 v1.9 版本中已成为 GA 版本。为了解决有状态服务的问题，它所管理的 Pod 拥有固定的 Pod 名称、启停顺序，在 StatefulSet 中，Pod 名字称为网络标识（hostname），还必须用到共享存储。

Deployment 中的服务是 service，而 StatefulSet 中的服务是 headless service，即无头服务，与 service 的区别就是后者没有 Cluster IP，解析其名称时将返回该 headless service 对应的全部 Pod 的 Endpoint 列表。

1.15　ConfigMap 概念剖析

在企业生产环境中，经常会遇到需要修改配置文件的情况。例如，Web 网站连接数据库的配置、业务系统之间相互调用配置、Kubernetes Pod 配置信息等，如果使用传统手动方式修改，则不仅会影响到服务的正常运行，操作步骤也很烦琐。

对于传统的应用服务而言，每个服务都有自己的配置文件，各自配置文件存储在服务所在的节点。对于单体应用，这种存储没有任何问题，但是随着用户数量的激增，一个节点不能满足线上用户的使用，故服务可能从一个节点扩展到十个节点，这就导致，如果有一个配置出现变更，就需要对应修改十次配置文件。

这种人工处理方式显然不能满足线上部署要求，故引入了各种类似于 ZooKeeper 中间件实现的配置中心，但配置中心属于"侵入式"设计，需要修改引入第三方类库，它要求每个业务都调用特定的配置接口，破坏了系统本身的完整性。

Kubernetes 利用 volume 功能完整设计了一套配置中心，其核心对象就是 ConfigMap，使用过程中不用修改任何原有设计，即可无缝对接 ConfigMap。

Kubernetes 项目从 1.2.0 版本开始引入了 ConfigMap 功能，主要用于将应用的配置信息与程序分离。这种方式不仅可以实现应用程序被复用，还可以通过不同的配置实现更灵活的功能。

在创建容器时，用户可以将应用程序打包为容器镜像，通过环境变量或外接挂载文件的方式进行配置注入。ConfigMap 非常灵活，相当于把配置文件信息单独存储在某处，需要时直接引用、挂载即可。

ConfigMap 以 key:value（K-V）的形式保存配置项，既可以表示一个变量的值（例如，

config=info），也可以表示一个完整配置文件的内容（例如，server.xml=<?xml…>…）。ConfigMap 在容器中使用的典型用法如下：

（1）将配置项设置为容器内的环境变量。

（2）将启动参数设置为环境变量。

（3）以 volume 的形式挂载到容器内部的文件或目录。

可以基于 kubectl create 指令创建 ConfigMap。例如，使用 ConfigMap 存储 MySQL 数据库的 IP 地址和端口信息，如图 1-5 所示。

```
kubectl create configmap mysql-config --from-literal=db.host=192.168.1.111
--from-literal=db.port=3306
kubectl get configmap mysql-config -o yaml
```

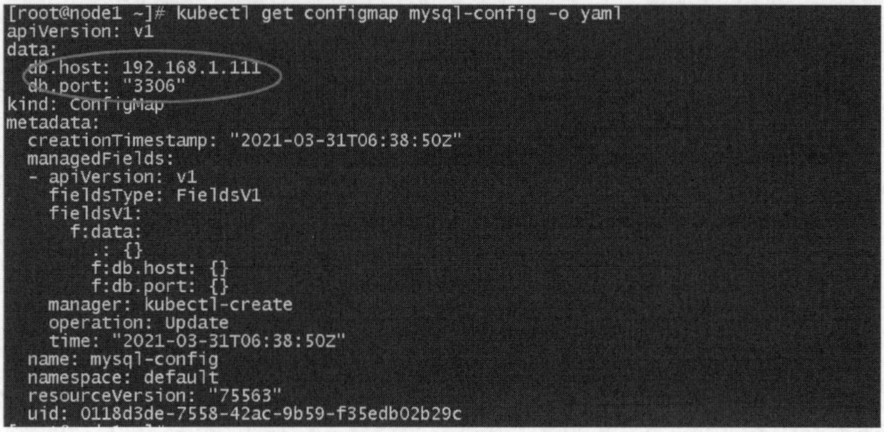

图 1-5　Kubernetes ConfigMap 配置界面

1.16　Secrets 概念剖析

在 Kubernetes 集群中，Secrets 通常用于存储和管理一些敏感数据，如密码、token、密钥等敏感信息。它把 Pod 想要访问的加密数据存放到 etcd 中。

用户可以通过在 Pod 容器里挂载 volume 或环境变量的方式访问这些 Secret 里保存的信息。Secret 有以下 3 种类型。

（1）Opaque：用来存储密码、密钥等，采用 base64 编码格式。但数据也可以通过 base64-decode 解码得到原始数据，加密性很弱。

（2）Service Account：用来访问 Kubernetes API，由 Kubernetes 自动创建，会自动挂载到 Pod

的 /run/secrets/kubernetes.io/serviceaccount 目录中。

（3）kubernetes.io/dockerconfigjson：用来存储私有 Docker registry 的认证信息。

1.17　CronJob 概念剖析

CronJob 用于创建基于时间调度的任务（Jobs）。一个 CronJob 对象就像 Crontab (Cron table) 文件中的一行。它用 Cron 格式进行编写，并周期性地在给定的调度时间执行。所有 CronJob 的 schedule 都是基于 kube-controller-manager 的时区。

如果 Kubernetes 在 Pod 或裸容器中运行了 kube-controller-manager，那么为该容器所设置的时区将会决定 CronJob 的控制器所使用的时区。

为 CronJob 资源创建清单时，请确保所提供的名称是一个合法的 DNS 子域名，名称不能超过 52 个字符。这是因为 CronJob 控制器将自动在提供的任务名称后附加 11 个字符，且存在一个限制，即任务名称的最大长度不能超过 63 个字符。

CronJob 对于创建周期性的、多次重复的任务很有用。例如，执行数据备份或者发送邮件。CronJob 也可以计划在指定时间执行的独立任务。例如，计划当集群看起来很空闲时执行某个任务。

（1）CronJob 案例会在每分钟打印出当前时间和问候消息，代码如下，如图 1-6 所示。

```
apiVersion: batch/v1beta1
kind: CronJob
metadata:
  name: hello
spec:
  schedule: "*/1 * * * *"
  jobTemplate:
    spec:
      template:
        spec:
          containers:
          - name: hello
            image: busybox
            imagePullPolicy: IfNotPresent
            command:
            - /bin/sh
            - -c
```

```
      - date; echo Hello from the Kubernetes cluster
      restartPolicy: OnFailure
```

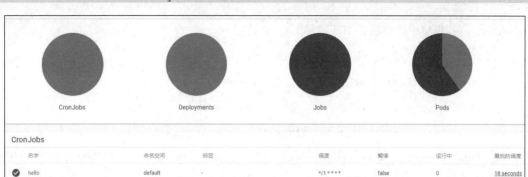

图 1-6　Kubernetes CronJob 任务创建

（2）CronJob 时间表语法如下所示。

```
#  ┌───────────────── 分 (0 ~ 59)
#  │ ┌─────────────── 时 (0 ~23)
#  │ │ ┌───────────── 日 (1 ~31)
#  │ │ │ ┌─────────── 月 (1 ~12)
#  │ │ │ │ ┌───────── 星期 (0 ~6)（星期日到星期一；在某些系统上,
#  7 也是星期日）
#  │ │ │ │ │
#  │ │ │ │ │
#  │ │ │ │ │
#  * * * * *
```

1.18　Kubernetes 证书剖析和制作实战

　　Kubernetes 需要公钥基础设施（Public Key Infrastructure，PKI）证书才能进行基于安全传输层协议（Transport Layer Security，TLS，主要用于在两个通信应用程序之间提供保密性和数据完整性）的身份验证。如果用户是通过 kubeadm 安装的 Kubernetes，则会自动生成集群所需的证书。用户还可以生成自己的证书。例如，不将私钥存储在 API 服务器上，可以让私钥更加安全。如果是通过 kubeadm 安装的 Kubernetes，则所有证书都存放在 /etc/kubernetes/pki 目录下。本文所有相关的路径都是基于该路径的相对路径。

　　PKI 采用证书进行公钥管理，通过第三方的可信任机构（认证中心，即 CA），把用户的公钥和其他标识信息捆绑在一起，其中包括用户名和电子邮件地址等信息，以在互联网上验证用户的身份。PKI 把公钥密码和对称密码结合起来，在互联网上实现密钥的自动管理，保证网上数据的安全传输。

Kubernetes 集群中需要的证书环节如下：

（1）etcd 对外提供服务，需要一套 etcd server 证书。

（2）etcd 各节点之间进行通信，需要一套 etcd peer 证书。

（3）kube-apiserver 访问 etcd，需要一套 etcd client 证书。

（4）kube-apiserver 对外提供服务，需要一套 kube-apiserver server 证书。

（5）kube-scheduler、kube-controller-manager、kube-proxy、kubelet 和其他可能用到的组件，需要访问 Kube-apiserver，需要一套 kube-apiserver client 证书。

（6）kube-controller-manager 要生成服务的 service account，需要一套用来签署 service account 的证书（CA 证书）。

（7）kubelet 对外提供服务，需要一套 kubelet server 证书。

（8）kube-apiserver 需要访问 kubelet，需要一套 kubelet client 证书。

同一个套里的证书必须是用同一个 CA 签署的，签署不同套里的证书的 CA 可以相同，也可以不同。例如，所有 etcd server 证书需要是同一个 CA 签署的，所有的 etcd peer 证书也是同一个 CA 签署的，而一个 etcd server 证书和一个 etcd peer 证书完全可以是两个 CA 签署的，彼此没有任何关系，这里就算是两套证书。

虽然可以用多套证书，但是维护多套 CA 实在过于繁杂，这里还是用一个 CA 签署所有证书。

（1）需要准备的证书如下：

```
admin.pem
ca.-key.pem
ca.pem
admin-key.pem
kube-scheduler-key.pem
kube-scheduler.pem
kube-controller-manager-key.pem
kube-controller-manager.pem
kube-proxy-key.pem
kube-proxy.pem
kubernetes-key.pem
kubernetes.pem
```

（2）使用证书的组件如下：

① etcd：ca.pem、kubernetes-key.pem、kubernetes.pem。

② Kube-apiserver：ca.pem、ca-key.pem、kubernetes-key.pem、kubernetes.pem。

③ Kubelet：ca.pem。

④ kube-proxy：ca.pem、kube-proxy-key.pem、kube-proxy.pem。

⑤ kubectl：ca.pem、admin-key.pem、admin.pem。

⑥ kube-controller-manager：ca-key.pem、ca.pem、kube-controller-manager-key.pem、kube-controller-manager.pem。

⑦ kube-scheduler：kube-scheduler-key.pem、kube-scheduler.pem。

此处使用 CFSSL 制作证书，CFSSL 是 Cloudflare 开发的一个开源的 PKI 工具，是一个完备的 CA 服务系统，可以签署、撤销证书等，覆盖了一个证书的整个生命周期。后面只用到了它的命令行工具。

注：一般情况下，Kubernetes 中的证书只需要创建一次，以后在向集群中添加新节点时只要将/etc/kubernetes/ssl 目录下的证书复制到新节点上即可。

```
wget https://pkg.cfssl.org/R1.2/cfssl_linux-amd64
wget https://pkg.cfssl.org/R1.2/cfssljson_linux-amd64
wget https://pkg.cfssl.org/R1.2/cfssl-certinfo_linux-amd64
chmod +x cfssl_linux-amd64 cfssljson_linux-amd64 cfssl-certinfo_linux-amd64
mv cfssl_linux-amd64 /usr/local/bin/cfssl
mv cfssljson_linux-amd64 /usr/local/bin/cfssljson
mv cfssl-certinfo_linux-amd64 /usr/bin/cfssl-certinfo
```

（3）创建 CA 证书，配置文件，操作指令如下：

```
cat>ca-config.json<<EOF
{
  "signing": {
    "default": {
      "expiry": "87600h"
    },
    "profiles": {
      "kubernetes": {
        "usages": [
            "signing",
            "key encipherment",
            "server auth",
```

```
            "client auth"
          ],
          "expiry": "87600h"
        }
      }
    }
}
EOF
```

(4) Kubernetes 证书配置文件参数剖析。

① ca-config.json：可以定义多个 profiles，指定不同的过期时间、使用场景等参数。后续在签名证书时使用某个 profile。

② signing：可以签名其他证书，生成的 ca.pem 证书中 CA=TRUE。

③ server auth：client 可以用该 CA 对 server 提供的证书进行验证。

④ client auth：server 可以用该 CA 对 client 提供的证书进行验证。

⑤ expiry：设置证书过期时间。

(5) 创建 CA 证书签名请求文件，操作指令如下：

```
cat>ca-csr.json<<EOF
{
  "CN": "kubernetes",
  "key": {
    "algo": "rsa",
    "size": 2048
  },
  "names": [
    {
      "C": "CN",
      "ST": "BeiJing",
      "L": "BeiJing",
      "O": "Kubernetes",
      "OU": "System"
    }
  ],
    "ca": {
       "expiry": "87600h"
    }
}
EOF
```

（6）CA 证书签名请求文件配置参数剖析。

① CN：Common Name，Kube-apiserver 从证书中提取该字段作为请求的用户名（User Name）。浏览器使用该字段验证网站是否合法。

② O：Organization，Kube-apiserver 从证书中提取该字段作为请求用户所属的组（Group）。

（7）生成 CA 证书和私钥，操作指令如下：

```
cfssl gencert -initca ca-csr.json | cfssljson -bare ca
ls | grep ca
ca-config.json
ca.csr
ca-csr.json
ca-key.pem
ca.pem
```

其中，ca-key.pem 是 CA 的私钥；ca.csr 是一个签署请求；ca.pem 是 CA 证书，是后面 Kubernetes 组件会用到的 RootCA。

（8）创建 Kubernetes 证书，并创建 Kubernetes 证书签名请求文件，操作指令如下：

```
cat>kubernetes-csr.json<<EOF
{
    "CN": "kubernetes",
    "hosts": [
      "127.0.0.1",
      "192.168.1.145",
      "192.168.1.146",
      "etcd01",
      "kubernetes",
      "kube-api.jd.com",
      "kubernetes.default",
      "kubernetes.default.svc",
      "kubernetes.default.svc.cluster",
      "kubernetes.default.svc.cluster.local"
    ],
    "key": {
        "algo": "rsa",
        "size": 2048
    },
    "names": [
        {
            "C": "CN",
```

```
        "ST": "BeiJing",
        "L": "BeiJing",
        "O": "Kubernetes",
        "OU": "System"
      }
    ]
}
EOF
```

如果 hosts 字段不为空，则需要指定授权使用该证书的 IP 地址或域名列表。由于该证书后续被 etcd 集群和 Kubernetes master 集群使用，将 etcd、master 节点的 IP 地址都填上，同时还有 service 网络的首 IP 地址（一般是 Kube-apiserver 指定的 service-cluster-ip-range 网段的第一个 IP 地址，如 10.0.0.1）。以上物理节点的 IP 地址也可以更换为主机名。

（9）生成 Kubernetes 证书和私钥，操作指令如下：

```
cfssl gencert -ca=ca.pem -ca-key=ca-key.pem -config=ca-config.json -profile=kubernetes kubernetes-csr.json | cfssljson -bare kubernetes
ls |grep kubernetes
kubernetes.csr
kubernetes-csr.json
kubernetes-key.pem
kubernetes.pem
```

（10）创建 admin 证书，并创建 admin 证书签名请求文件，操作指令如下：

```
cat>admin-csr.json<<EOF
{
  "CN": "admin",
  "hosts": [],
  "key": {
    "algo": "rsa",
    "size": 2048
  },
  "names": [
    {
      "C": "CN",
      "ST": "BeiJing",
      "L": "BeiJing",
      "O": "system:masters",
      "OU": "System"
    }
  ]
```

```
}
EOF
```

Kube-apiserver 使用 RBAC 对客户端（如 kubelet、kube-proxy、Pod）请求进行授权。Kube-apiserver 预定义了一些 RBAC 使用的 RoleBindings。例如，cluster-admin 将 Group system:masters 与 Role cluster-admin 绑定，该 Role 授予了调用 Kube-apiserver 的所有 API 的权限。

O 指定该证书的 Group 为 system:masters，Kubelet 使用该证书访问 Kube-apiserver 时，由于证书被 CA 签名，所以认证通过，同时由于证书用户组为经过预授权的 system:masters，所以被授予访问所有 API 的权限。

注：这个 admin 证书用于生成管理员用的 kube config 配置文件，一般建议使用 RBAC 对 Kubernetes 进行角色权限控制，Kubernetes 将证书中的 CN 字段作为 User，O 字段作为 Group。

（11）生成 admin 证书和私钥，操作指令如下：

```
cfssl gencert -ca=ca.pem -ca-key=ca-key.pem -config=ca-config.json -profile=kubernetes admin-csr.json | cfssljson -bare admin
ls | grep admin
admin.csr
admin-csr.json
admin-key.pem
admin.pem
```

（12）创建 kube-proxy 证书及 kube-proxy 证书签名请求文件，操作指令如下：

```
cat>kube-proxy-csr.json<<EOF
{
  "CN": "system:kube-proxy",
  "hosts": [],
  "key": {
    "algo": "rsa",
    "size": 2048
  },
  "names": [
    {
      "C": "CN",
      "ST": "BeiJing",
      "L": "BeiJing",
      "O": "Kubernetes",
      "OU": "System"
    }
```

```
    ]
}
EOF
```

Kube-apiserver 预定义的 RoleBinding system:node-proxier 将 User system:kube-proxy 与 Role system:node-proxier 绑定，该 Role 授予了调用 Kube-apiserver Proxy 相关 API 的权限。CN 指定证书的 User 为 system:kube-proxy。

因为该证书只会被 kubectl 当作 client 证书使用，所以 hosts 字段为空，生成 kube-proxy 证书和私钥。

```
cfssl gencert -ca=ca.pem -ca-key=ca-key.pem -config=ca-config.json -profile=kubernetes  kube-proxy-csr.json | cfssljson -bare kube-proxy
ls |grep kube-proxy
kube-proxy.csr
kube-proxy-csr.json
kube-proxy-key.pem
kube-proxy.pem
```

（13）创建 kube-controller-manager 证书，并创建其证书签名请求文件，操作指令如下：

```
cat>kube-controller-manager-csr.json<<EOF
{
    "CN": "system:kube-controller-manager",
    "key": {
        "algo": "rsa",
        "size": 2048
    },
    "hosts": [
      "127.0.0.1",
      "192.168.1.145",
      "192.168.1.146",
      "Kubernetes-master1",
    ],
    "names": [
      {
        "C": "CN",
        "ST": "BeiJing",
        "L": "BeiJing",
        "O": "system:kube-controller-manager",
        "OU": "system"
      }
```

```
    ]
}
EOF
```

其中，hosts 列表包含所有 kube-controller-manager 节点 IP；CN 为 system:kube-controller-manager；O 为 system:kube-controller-manager。

（14）生成 kube-controller-manager 证书和私钥，操作指令如下：

```
cfssl gencert -ca=ca.pem -ca-key=ca-key.pem -config=ca-config.json -profile=kubernetes kube-controller-manager-csr.json | cfssljson -bare kube-controller-manager
```

（15）创建 kube-scheduler 证书签名请求文件，操作指令如下：

```
cat>kube-scheduler-csr.json<<EOF
{
    "CN": "system:kube-scheduler",
    "hosts": [
      "127.0.0.1",
      "192.168.1.145",
      "192.168.1.146",
      "Kubernetes-master1",
    ],
    "key": {
        "algo": "rsa",
        "size": 2048
    },
    "names": [
      {
        "C": "CN",
        "ST": "BeiJing",
        "L": "BeiJing",
        "O": "system:kube-scheduler",
        "OU": "4Paradigm"
      }
    ]
}
EOF
```

（16）根据以上创建的 CA 证书和 JSON 文件，接下来创建 Kubernetes 各个服务所需证书，操作指令如下：

```
cfssl gencert -ca=ca.pem -ca-key=ca-key.pem -config=ca-config.json -
profile=kubernetes kube-scheduler-csr.json| cfssljson -bare kube-scheduler
ls | grep pem
admin-key.pem
admin.pem
ca-key.pem
ca.pem
kube-proxy-key.pem
kube-proxy.pem
kubernetes-key.pem
kubernetes.pem
kube-controller-manager-key.pem
kube-controller-manager.pem
kube-scheduler-key.pem
kube-scheduler.pem
```

(17)查看证书信息,操作指令如下:

```
cfssl-certinfo -cert kubernetes.pem
```

部署 Kubernetes 云计算平台时,将这些证书文件分发至此集群中其他节点机器对应目录即可。至此,TLS 证书创建完毕。

第 2 章 Kubernetes 云计算平台配置实战

部署 Kubernetes 云计算平台，至少准备 2 台服务器，此处为 4 台，包括 1 台 Docker 仓库。

Kubernetes master1 节点：192.168.1.145
Kubernetes node1 节点：192.168.1.146
Kubernetes node2 节点：192.168.1.147
Docker 私有库节点：192.168.1.148

2.1 Kubernetes 节点 hosts 及防火墙设置

Kubernetes master1、node1、node2 节点均执行如下代码，设置 hosts 和防火墙配置，操作指令如下：

```
#添加 hosts 解析
cat >/etc/hosts<<EOF
127.0.0.1 localhost localhost.localdomain
192.168.1.145 master1
192.168.1.146 node1
192.168.1.147 node2
EOF
#临时关闭 SELinux 和防火墙
sed -i '/SELINUX/s/enforcing/disabled/g'  /etc/sysconfig/selinux
setenforce  0
systemctl   stop     firewalld.service
systemctl   disable    firewalld.service
#同步节点时间
yum install ntpdate -y
ntpdate pool.ntp.org
```

```
#修改对应节点主机名
hostname `cat /etc/hosts|grep $(ifconfig|grep broadcast|awk '{print
$2}')|awk '{print $2}'`;su
#关闭swapoff
swapoff -a
```

2.2　Linux 内核参数设置和优化

Kubernetes master1、node1、node2 节点均执行如下代码，设置 IPVS 模式并调整内核参数，操作指令如下：

```
cat > /etc/modules-load.d/ipvs.conf <<EOF
#Load IPVS at boot
ip_vs
ip_vs_rr
ip_vs_wrr
ip_vs_sh
nf_conntrack_ipv4
EOF
systemctl enable --now systemd-modules-load.service
#确认内核模块加载成功
lsmod | grep -e ip_vs -e nf_conntrack_ipv4
#安装ipset、ipvsadm
yum install -y ipset ipvsadm
#配置内核参数
cat <<EOF > /etc/sysctl.d/Kubernetes.conf
net.bridge.bridge-nf-call-ip6tables=1
net.bridge.bridge-nf-call-iptables=1
EOF
sysctl --system
```

2.3　Docker 虚拟化案例实战

Kubernetes master1、node1、node2 节点均执行如下代码，部署 Docker 虚拟化软件，操作指令如下：

```
#安装依赖软件包
yum install -y yum-utils device-mapper-persistent-data lvm2
```

```
#添加Docker repository,这里使用国内阿里云yum源
yum-config-manager   --add-repo http://mirrors.aliyun.com/docker-ce/linux/centos/docker-ce.repo
#安装docker-ce,这里直接安装最新版本
yum install -y docker-ce
#修改Docker配置文件
mkdir /etc/docker
cat > /etc/docker/daemon.json <<EOF
{
  "exec-opts": ["native.cgroupdriver=systemd"],
  "log-driver": "json-file",
  "log-opts": {
    "max-size": "100m"
  },
  "storage-driver": "overlay2",
  "storage-opts": [
    "overlay2.override_kernel_check=true"
  ],
  "registry-mirrors": ["https://uyah70su.mirror.aliyuncs.com"]
}
EOF
#注意,由于国内拉取镜像较慢,配置文件最后增加了代码registry-mirrors
#mkdir -p /etc/systemd/system/docker.service.d
#重启Docker服务
systemctl daemon-reload
systemctl enable docker.service
systemctl start docker.service
ps -ef|grep -aiE docker
```

2.4 Kubernetes 添加部署源

Kubernetes master1、node1、node2 节点均执行如下代码,添加 Kubernetes 网络安装源,操作指令如下:

```
cat>>/etc/yum.repos.d/kubernetes.repo<<EOF
[kubernetes]
name=Kubernetes
baseurl=https://mirrors.aliyun.com/kubernetes/yum/repos/kubernetes-el7-x86_64
enabled=1
```

```
gpgcheck=0
repo_gpgcheck=0
gpgkey=https://mirrors.aliyun.com/kubernetes/yum/doc/yum-key.gpg
EOF
```

2.5　Kubernetes Kubeadm 案例实战

Kubeadm 是一个 Kubernetes 部署工具，它提供了 kubeadm init 及 kubeadm join 这两个命令快速创建 Kubernetes 集群。

Kubeadm 通过执行必要的操作启动和运行一个最小可用的集群。它被设计为只关心启动集群，而不是之前的节点准备工作。诸如安装各种各样值得拥有的插件，如 Kubernetes Dashboard、监控解决方案以及特定云提供商的插件，这些都不在 Kubeadm 负责的范围内。

相反，我们期望由一个基于 Kubeadm、从更高层设计的、更加合适的工具来做这些事情；并且，理想情况下，使用 Kubeadm 作为所有部署的基础将会使创建一个符合期望的集群变得容易。

Kubeadm 是用于初始化 Kubernetes Cluster 的工具。kubelet 运行在 Cluster 所有节点上，负责调用 Docker 指令，启动 Pod 和容器。kubectl:kubectl 是 Kubernetes 命令行工具，通过 kubectl 可以部署和管理应用，查看各种资源，创建、删除和更新组件。

在计算机上手动安装 Docker、Kubeadm、kubelet、kubectl 几个二进制文件，然后才能再容器化部署其他 Kubernetes 组件。主要通过 Kubeadm init 初始化，如图 2-1 所示。初始化 Kubernetes 集群的流程如下。

（1）检查工作（Preflight Checks）：检查 Linux 内核版本、Cgroups 模块可用性、组件版本、端口占用情况、Docker 等的依赖情况。

（2）生成对外提供服务的 CA 证书及对应的目录。

（3）生成其他组件访问 kube-apiserver 所需的配置文件。

（4）为 master 组件生成 Pod 配置文件，利用这些配置文件，通过 Kubernetes 中特殊的容器启动方法 Static Pod（kubelet 启动时自动加载固定目录的 Pod YAML 文件并启动）可以以 Pod 方式部署 kube-apiserver、kube-controller-manager、kube-scheduler 三个 master 组件。同时还会生成 etcd 的 Pod YAML 文件。

（5）为 master 节点打标签。

（6）为集群生成一个 Bootstrap token，其他节点加入集群的计算机和 API server 打交道，需要获取相应的证书文件，所以 Bootstrap token 需要扮演安全验证的角色。

（7）生成 node 节点加入集群时要使用的其他必要配置。

（8）安装默认插件，如 kube-proxy 和 Core DNS，分别提供集群的服务发现和 DNS 功能。

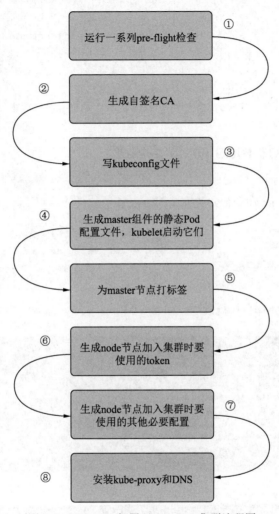

图 2-1　Kubeadm 部署 Kubernetes 集群流程图

安装 Kubeadm 工具，操作指令如下：

```
#安装 Kubeadm
yum install -y kubeadm-1.20.4 kubelet-1.20.4 kubectl-1.20.4
#启动 kubelet 服务
systemctl enable kubelet.service
systemctl start kubelet.service
```

Kubeadm 常见指令操作详解如下：

```
kubeadm init         #用于启动一个 Kubernetes 主节点
kubeadm join         #用于启动一个 Kubernetes 工作节点并将其加入集群
kubeadm upgrade      #用于更新一个 Kubernetes 集群到新版本
kubeadm config       #如果使用 v1.7.x 或更低版本的 kubeadm 初始化集群，则需要对集
                     #群做一些配置以便使用 kubeadm upgrade 命令
kubeadm token        #用于管理 kubeadm join 使用的令牌
kubeadm reset        #用于还原 kubeadm init 或者 kubeadm join 对主机所做的任何更改
kubeadm version      #用于打印 kubeadm 版本
kubeadm alpha        #用于预览一组可用的新功能，以便从社区搜集反馈
```

2.6　Kubernetes master 节点实战

（1）执行 kubeadm init 命令初始化安装 Master 相关软件。

```
kubeadm init    --control-plane-endpoint=192.168.1.145:6443 --image-repository registry.aliyuncs.com/google_containers    --kubernetes-version v1.20.4    --service-cidr=10.10.0.0/16    --pod-network-cidr=10.244.0.0/16 --upload-certs
```

（2）根据以上指令操作，执行成功，如图 2-2 所示。

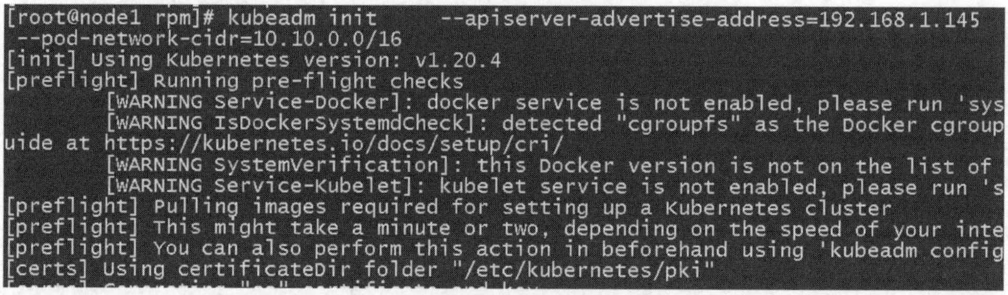

（a）

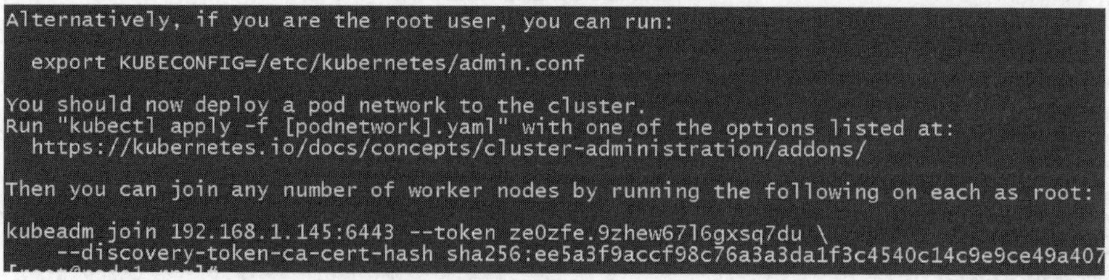

（b）

图 2-2　Kubeadm 初始化 Kubernetes 集群

（3）根据图 2-2 提示，接下来需手动执行如下指令，复制 admin 配置文件，操作指令如下：

```
mkdir -p $HOME/.kube
sudo cp -i /etc/kubernetes/admin.conf $HOME/.kube/config
sudo chown $(id -u):$(id -g) $HOME/.kube/config
```

（4）将 master 节点初始化之后，要将 node 节点也加入 Kubernetes 集群，操作指令如下：

```
kubeadm join 192.168.1.145:6443 --token ze0zfe.9zhew67l6gxsq7du \
    --discovery-token-ca-cert-hash
sha256:ee5a3f9accf98c76a3a3da1f3c4540c14c9e9ce49a4070de4b832aa8cb3a8f31
```

2.7 Kubernetes 集群节点和删除

根据 2.6 节中的所有操作和步骤，Kubernetes 集群部署成功，如果要将 node2 节点也加入 Kubernetes 集群，需要执行的操作指令如下：

```
#启动 node1 节点上的 docker 引擎服务
systemctl start docker.service
#将 node1 节点加入 Kubernetes 集群
kubeadm join 192.168.1.145:6443 --token ze0zfe.9zhew67l6gxsq7du \
    --discovery-token-ca-cert-hash sha256:ee5a3f9accf98c76a3a3da1f3c4540c14c9e9ce49a4070de4b832aa8cb3a8f31
#执行命令 kubeadm init 时没有记录下加入集群的指令,可以通过以下命令重新创建
kubeadm token create --print-join-command
#登录 Kubernetes master 节点验证节点信息
kubectl get nodes
#删除一个节点前,先驱赶掉上面的 Pod 容器
kubectl drain node2 --delete-local-data --force --ignore-daemonsets
#在 master 端执行如下指令
kubectl delete node/node2
```

2.8 Kubernetes 节点网络配置

Kubernetes 整个集群所有服务器（master、Minion）配置 Flannel，必须安装 Pod 网络插件，以便 Pod 之间相互通信。必须在全部应用程序之前部署网络，CoreDNS 不会在安装网络插件之前启动。

（1）安装 Flannel 网络插件。Flannel 定义 Pod 的网段为 10.244.0.0/16，Pod 容器的 IP 地址会

自动分配 10.244 开头的网段 IP。部署 Flannel 网络插件操作指令如下:

```
#下载 Flannel 插件 YML 文件
yum install wget -y
wget https://raw.githubusercontent.com/coreos/flannel/master/Documentation/kube-flannel.yml
#提前下载 Flanneld 组建所需镜像
for i in $(cat kube-flannel.yml |grep image|awk -F: '{print $2":"$3}' |uniq );do docker pull $i ;done
#应用 YML 文件
kubectl apply -f kube-flannel.yml
```

（2）如果 Kube-flannel.yml 配置文件无法下载，也可以将以下代码直接复制并且写入 Kube-flannel.yml。文件代码如下:

```
---
apiVersion: policy/v1beta1
kind: PodSecurityPolicy
metadata:
  name: psp.flannel.unprivileged
  annotations:
    seccomp.security.alpha.kubernetes.io/allowedProfileNames: docker/default
    seccomp.security.alpha.kubernetes.io/defaultProfileName: docker/default
    apparmor.security.beta.kubernetes.io/allowedProfileNames: runtime/default
    apparmor.security.beta.kubernetes.io/defaultProfileName: runtime/default
spec:
  privileged: false
  volumes:
  - configMap
  - secret
  - emptyDir
  - hostPath
  allowedHostPaths:
  - pathPrefix: "/etc/cni/net.d"
  - pathPrefix: "/etc/kube-flannel"
  - pathPrefix: "/run/flannel"
  readOnlyRootFilesystem: false
  #Users and groups
```

```yaml
    runAsUser:
      rule: RunAsAny
    supplementalGroups:
      rule: RunAsAny
    fsGroup:
      rule: RunAsAny
    #Privilege Escalation
    allowPrivilegeEscalation: false
    defaultAllowPrivilegeEscalation: false
    #Capabilities
    allowedCapabilities: ['NET_ADMIN', 'NET_RAW']
    defaultAddCapabilities: []
    requiredDropCapabilities: []
    #Host namespaces
    hostPID: false
    hostIPC: false
    hostNetwork: true
    hostPorts:
    - min: 0
      max: 65535
    # seLinux
    seLinux:
      # seLinux is unused in CaaSP
      rule: 'RunAsAny'
---
kind: ClusterRole
apiVersion: rbac.authorization.Kubernetes.io/v1
metadata:
  name: flannel
rules:
- apiGroups: ['extensions']
  resources: ['podsecuritypolicies']
  verbs: ['use']
  resourceNames: ['psp.flannel.unprivileged']
- apiGroups:
  - ""
  resources:
  - pods
  verbs:
  - get
- apiGroups:
```

```
    - ""
      resources:
      - nodes
      verbs:
      - list
      - watch
    - apiGroups:
      - ""
      resources:
      - nodes/status
      verbs:
      - patch
---
kind: ClusterRoleBinding
apiVersion: rbac.authorization.Kubernetes.io/v1
metadata:
  name: flannel
roleRef:
  apiGroup: rbac.authorization.Kubernetes.io
  kind: ClusterRole
  name: flannel
subjects:
- kind: ServiceAccount
  name: flannel
  namespace: kube-system
---
apiVersion: v1
kind: ServiceAccount
metadata:
  name: flannel
  namespace: kube-system
---
kind: ConfigMap
apiVersion: v1
metadata:
  name: kube-flannel-cfg
  namespace: kube-system
  labels:
    tier: node
    app: flannel
data:
```

```yaml
  cni-conf.json: |
    {
      "name": "cbr0",
      "cniVersion": "0.3.1",
      "plugins": [
        {
          "type": "flannel",
          "delegate": {
            "hairpinMode": true,
            "isDefaultGateway": true
          }
        },
        {
          "type": "portmap",
          "capabilities": {
            "portMappings": true
          }
        }
      ]
    }
  net-conf.json: |
    {
      "Network": "10.244.0.0/16",
      "Backend": {
        "Type": "vxlan"
      }
    }
---
apiVersion: apps/v1
kind: DaemonSet
metadata:
  name: kube-flannel-ds
  namespace: kube-system
  labels:
    tier: node
    app: flannel
spec:
  selector:
    matchLabels:
      app: flannel
  template:
```

```yaml
metadata:
  labels:
    tier: node
    app: flannel
spec:
  affinity:
    nodeAffinity:
      requiredDuringSchedulingIgnoredDuringExecution:
        nodeSelectorTerms:
        - matchExpressions:
          - key: kubernetes.io/os
            operator: In
            values:
            - linux
  hostNetwork: true
  priorityClassName: system-node-critical
  tolerations:
  - operator: Exists
    effect: NoSchedule
  serviceAccountName: flannel
  initContainers:
  - name: install-cni
    image: quay.io/coreos/flannel:v0.13.1-rc2
    command:
    - cp
    args:
    - -f
    - /etc/kube-flannel/cni-conf.json
    - /etc/cni/net.d/10-flannel.conflist
    volumeMounts:
    - name: cni
      mountPath: /etc/cni/net.d
    - name: flannel-cfg
      mountPath: /etc/kube-flannel/
  containers:
  - name: kube-flannel
    image: quay.io/coreos/flannel:v0.13.1-rc2
    command:
    - /opt/bin/flanneld
    args:
    - --ip-masq
```

```yaml
        - --kube-subnet-mgr
        resources:
          requests:
            cpu: "100m"
            memory: "50Mi"
          limits:
            cpu: "100m"
            memory: "50Mi"
        securityContext:
          privileged: false
          capabilities:
            add: ["NET_ADMIN", "NET_RAW"]
        env:
        - name: POD_NAME
          valueFrom:
            fieldRef:
              fieldPath: metadata.name
        - name: POD_NAMESPACE
          valueFrom:
            fieldRef:
              fieldPath: metadata.namespace
        volumeMounts:
        - name: run
          mountPath: /run/flannel
        - name: flannel-cfg
          mountPath: /etc/kube-flannel/
      volumes:
      - name: run
        hostPath:
          path: /run/flannel
      - name: cni
        hostPath:
          path: /etc/cni/net.d
      - name: flannel-cfg
        configMap:
          name: kube-flannel-cfg
```

（3）查看 Kubernetes Flannel 网络插件是否部署成功，如图 2-3 所示，操作指令如下：

```
kubectl -n kube-system get pods|grep -aiE flannel
```

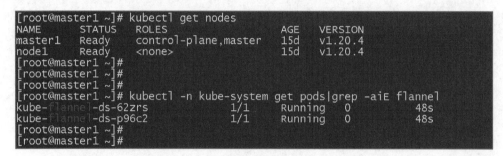

图 2-3　Kubernetes Flannel 网络插件部署

（4）安装 Calico 网络插件。

Kubernetes 云计算平台网络通信除了使用 Flannel 之外，还可以使用 Calico（一种容器之间互通的网络方案）。两个网络插件都可以实现 Kubernetes 容器互通。实际生产环境中二选一即可。

在虚拟化平台中（如 OpenStack、Docker 等）都需要实现容器之间互连，但同时也需要对容器进行隔离控制，就像在互联网中的服务仅开放 80 端口、公有云的多租户一样，提供隔离和管控机制。

在多数虚拟化平台实现中，通常都使用二层隔离技术实现容器的网络，这些技术有一些弊端。例如，需要依赖 VLAN、Bridge 和隧道等技术，其中 Bridge 带来了复杂性，VLAN 隔离和 Tunnel 隧道则消耗更多的资源并对物理环境有要求，随着网络规模增大，整体会变得越加复杂。

我们尝试把 Host 当作 Internet 中的路由器，同样使用 BGP 同步路由，并使用 iptables 来做安全访问策略，最终设计出了 Calico 方案。

Calico 不使用隧道或 NAT 来实现转发，而是巧妙地把所有二、三层流量转换成三层流量，并通过 Host 上路由配置完成跨 Host 转发。

为了保证 Calico 正常工作，需要传递--pod-network-cidr=10.10.0.0/16 到 kubeadm init 或更新 calico.yaml 文件，以与 Pod 网络相匹配。

（1）部署 Calico 网络插件，同样需要从官网下载 calico.yaml 文件，操作指令如下：

```
kubectl apply -f https://docs.projectcalico.org/v3.10/manifests/calico.yaml
```

（2）如果安装 Kubernetes Flannel 网络插件，必须通过 kubeadm init 配置 -pod-network-cidr=10.10.0.0/16 参数。

（3）验证 Calico 网络插件，安装 Calico Pod 网络后，确认 Coredns 及其他 Pod 全部运行正常，查看 master 节点状态为 Ready，如图 2-4 所示，操作指令如下：

```
kubectl get nodes
kubectl -n kube-system get pods
```

```
[root@node1 rpm]# kubectl get nodes
NAME    STATUS   ROLES                  AGE    VERSION
node1   Ready    control-plane,master   8m22s  v1.20.4
[root@node1 rpm]#
[root@node1 rpm]# kubectl -n kube-system get pods
NAME                                         READY   STATUS              RESTARTS
calico-kube-controllers-7854b85cf7-v8qhr     0/1     ContainerCreating   0
calico-node-swfn4                            0/1     PodInitializing     0
coredns-7f89b7bc75-2gs55                     0/1     ContainerCreating   0
coredns-7f89b7bc75-gqdg9                     0/1     ContainerCreating   0
etcd-node1                                   1/1     Running             0
kube-apiserver-node1                         1/1     Running             0
kube-controller-manager-node1                1/1     Running             0
```

图 2-4 Kubernetes Calico 网络插件部署

至此，Kubernetes 的 master1 节点和 node1 节点部署完成。接下来就可以管理和使用 Kubernetes 集群了。

2.9 Kubernetes 开启 IPVS 模式

IPVS（IP Virtual Server）实现了传输层负载均衡，也就是常说的 4 层 LAN 交换。Linux 内核默认集成 IPVS。IPVS 运行在主机上，在真实服务器集群前充当负载均衡器。

Kubernetes 默认使用 iptables 实现服务访问和代理，如果使用 IPVS 模式进行服务访问代理，kube-proxy 会监视 Kubernetes 集群中的对象和端点（endpoint），调用 netlink 接口以相应地创建 IPVS 规则并定期与 Kubernetes 中的 service 对象和 endpoints 同步 IPVS 规则，以确保 IPVS 状态与期望的一致。当访问 service 时，流量就会被重定向到后端的 Pod 上。

修改 kube-proxy 的 configmap，在 config.conf 中找到 mode 参数，改为 mode: "ipvs"并保存，操作指令如下：

```
kubectl -n kube-system get cm kube-proxy -o yaml | sed 's/mode: ""/mode: "ipvs"/g' | kubectl replace -f -
#或者手动修改
kubectl -n kube-system edit cm kube-proxy
kubectl -n kube-system get cm kube-proxy -o yaml | grep mode
    mode: "ipvs"
#重启 kube-proxy Pod
kubectl -n kube-system delete pods -l Kubernetes-app=kube-proxy
#确认 IPVS 模式开启成功
kubectl -n kube-system logs -f -l Kubernetes-app=kube-proxy | grep ipvs
```

```
#日志中打印出 Using ipvs Proxier,说明 IPVS 模式已经开启
#创建 service 之后,可以使用命令 ipvsadm -L -n 查看是否采用 IPVS 模式进行转发
ipvsadm -L -n
```

2.10 Kubernetes 集群故障排错

Kubernetes 集群可能出现 Pod 或 node 节点状态异常等情况,可以通过查看日志分析错误原因。常见排错操作方法和指令如下:

```
#查看 Pod 日志
kubectl -n kube-system logs -f <pod name>
#查看 Pod 运行状态及事件
kubectl -n kube-system describe pods <pod name>
node not ready                          #可以通过分析 node 相关日志排查
kubectl describe nodes <node name>
systemctl status docker
systemctl status kubelet
journalctl -xeu docker
journalctl -xeu kubelet
tail -f /var/log/messages
```

2.11 Kubernetes 集群节点移除

Kubernetes 云计算平台部署成功之后,在运行过程中,随着服务器时间寿命增加,会淘汰下架一些服务器,此时需要删除节点。以删除 node1 节点为例,在 master1 节点上执行如下指令:

```
kubectl drain node1 --delete-local-data --force --ignore-daemonsets
kubectl delete node node1
kubeadm reset -f
```

2.12 etcd 分布式案例操作

etcd 是一个分布式的、高可用的、一致的 key-value 存储数据库,基于 Go 语言实现,主要用于共享配置和服务发现。

在 Kubernetes 云计算平台中,Kubernetes 服务配置信息的管理共享和服务发现是一个很基本、很重要的问题。etcd 可以集中管理配置信息,服务端将配置信息存储于 etcd,客户端通过

etcd 得到服务配置信息，etcd 监听配置信息的改变，发现改变后通知客户端。

为了防止单点故障，还可以启动多个 etcd 组成集群。etcd 集群使用 raft 一致性算法处理日志复制，保证多节点数据的强一致性。

根据以上 Kubernetes 集群部署，etcd 默认也会被自动部署成功，可以使用以下指令测试 etcd 集群是否正常。

```
#查看etcd集群节点列表
etcdctl  member list
#查看etcd集群节点状态
etcdctl cluster-health
#获取etcd中config key的值
etcdctl get /atomic.io/network/config
#查看etcd目录树
etcdctl ls /atomic.io/network/subnets
#删除etcd目录树
etcdctl  rm   /atomic.io/network/   --recursive
#创建etcd config key 和 value
etcdctl  mk  /atomic.io/network/config '{"Network":"172.17.0.0/16"}'
```

Kubernetes 的 node 节点搭建和配置 Flannel 网络，etcd 中的/atomic.io/network/config 节点会被 node 节点上的 Flannel 用来创建 Docker IP 地址网段。查看 etcd 配置中心的网段信息，操作指令如下：

```
etcdctl ls /atomic.io/network/subnets
```

第 3 章 Kubernetes 企业网络 Flannel 实战

Flannel 是 CoreOS 团队针对 Kubernetes 设计的一个覆盖网络（Overlay Network）工具，其目的在于帮助每一个使用 Kuberentes 的 CoreOS 主机拥有一个完整的子网。

Flannel 通过给每台宿主机分配一个子网的方式为容器提供虚拟网络，它基于 Linux TUN/TAP，使用 UDP 封装 IP 包来创建 overlay 网络，并借助 etcd 维护网络的分配情况。

3.1 Flannel 工作原理

Flannel 是 CoreOS 团队针对 Kubernetes 设计的一个网络规划服务，简单来说，它的功能是让集群中的不同节点主机创建的 Docker 容器都具有全集群唯一的虚拟 IP 地址。

在默认的 Docker 配置中，每个 node 的 Docker 服务会分别负责所在节点容器的 IP 分配。node 内部的容器之间可以相互访问，但是跨主机网络相互间是不能通信的。

Flannel 的设计目的就是为集群中所有节点重新规划 IP 地址的使用规则，从而使得不同节点上的容器能够获得同属一个内网且不重复的 IP 地址，并让属于不同节点上的容器能够直接通过内网 IP 地址进行通信。

Flannel 使用 etcd 存储配置数据和子网分配信息。Flannel 启动之后，后台进程首先检索配置和正在使用的子网列表，然后选择一个可用的子网，并尝试注册它。etcd 也存储每个主机对应的 IP。

Flannel 使用 etcd 的 watch 机制监视/atomic.io/network/subnets 下面所有元素的变化信息，并且根据它来维护一个路由表。为了提高性能，Flannel 优化了 Universal TAP/TUN 设备，并对 TUN 和 UDP 之间的 IP 分片做了代理。

3.2 Flannel 架构介绍

Flannel 默认使用 8285 端口作为 UDP 封装报文的端口，VxLan 使用 8472 端口，如图 3-1 所示。

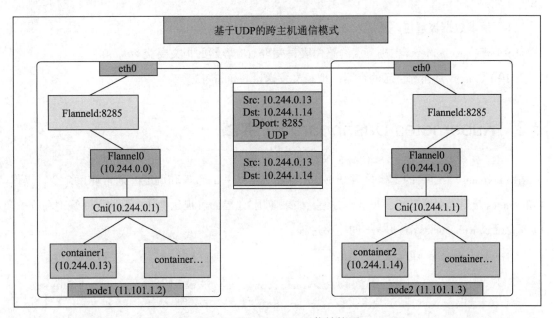

图 3-1　Flannel UDP 通信结构图

一个网络报文请求是怎么从一个容器发送到另外一个容器的呢？例如，从 master1 上的 container1 容器（IP 地址：10.244.0.13）访问 node1 上面的 container2 容器（IP 地址：10.244.1.14）。

（1）container1 容器 10.244.0.13 直接访问目标容器 container2 的 IP 地址 10.244.1.14，请求默认通过容器内部的 eth0 网卡发送出去。

（2）请求报文通过 Veth pair（虚拟设备对）被发送到 Docker 宿主机 VethXXX 设备上。

（3）VethXXX 设备是直接连接到虚拟交换机 Cni0 的，所以请求报文通过虚拟 Bridge Cni0 发送出去。

（4）查找 Docker 宿主机的路由表信息，同时外部容器 IP 的报文都会转发到 Flannel0 虚拟网卡（这是一个 P2P 的虚拟网卡），然后报文就被转发到监听在另一端的 flanneld 进程。

（5）因为 Flannel 在 etcd 中存储着子网和宿主机 IP 的对应关系，所以能够找到 10.244.1.14 对应的宿主机 IP 地址为 11.101.1.3，进而开始组装 UDP 数据包，并发送数据到目的主机。

（6）这个请求得以完成的原因是每个节点上都启动了一个 flanneld udp 进程，都监听着

8285 端口,所以 master1 通过 flanneld 进程把数据包通过宿主机的 Interface 网卡发送给 node1 的 flanneld 进程的相应端口即可。

(7)请求报文到达 node1 之后,继续往上传输到传输层,交给监听在 8285 端口的 flanneld 程序处理。

(8)请求数据被解包,然后发送给 Flannel0 虚拟网卡。

(9)查找 Kubernetes 主机节点的路由表,发现对应容器的报文要交给 Cni0。

(10)Cni0 找到连到自己的容器,把报文发送给 container2。

3.3 Kubernetes Dashboard UI 实战

Kubernetes 最重要的工作是对 Docker 容器集群统一的管理和调度。通常使用命令行操作 Kubernetes 集群及各个节点,非常不方便,如果使用 UI 界面可视化操作,更加方便管理和维护。以下为配置 Kubernetes dashboard 的完整过程。

(1)下载 dashboard 配置文件。

```
wget https://raw.githubusercontent.com/kubernetes/dashboard/v2.0.0-rc5/aio/deploy/recommended.yaml
\cp recommended.yaml recommended.yaml.bak
```

(2)修改文件 recommended.yaml 的 39 行内容。因为默认情况下 service 的类型是 cluster IP,所以需更改为 NodePort 的方式,便于访问,也可映射到指定的端口。

```
spec:
  type: NodePort
  ports:
    - port: 443
      targetPort: 8443
      nodePort: 31001
  selector:
    Kubernetes -app: kubernetes-dashboard
```

(3)修改文件 recommended.yaml 的 195 行内容。因为默认情况下 Dashboard 为英文显示,故可以设置为中文。

```
env:
        - name: ACCEPT_LANGUAGE
          value: zh
```

（4）创建 Dashboard 服务，操作指令如下：

```
kubectl apply -f recommended.yaml
```

（5）查看 Dashboard 运行状态。

```
kubectl get pod -n kubernetes-dashboard
kubectl get svc -n kubernetes-dashboard
```

（6）基于 Token 的方式访问，设置和绑定 Dashboard 权限，获取 Token 值，如图 3-2 所示，操作指令如下：

```
#创建 Dashboard 的管理用户
kubectl create serviceaccount dashboard-admin -n kube-system
#将创建的 Dashboard 用户绑定为管理用户
kubectl create clusterrolebinding dashboard-cluster-admin --clusterrole=cluster-admin --serviceaccount=kube-system:dashboard-admin
#获取刚刚创建的用户对应的 Token 名称
kubectl get secrets -n kube-system | grep dashboard
#查看 Token 的详细信息
kubectl describe secrets -n kube-system $(kubectl get secrets -n kube-system | grep dashboard |awk '{print $1}')
```

(a)

(b)

图 3-2 Kubernetes Token 密钥值获取

（7）通过浏览器访问 Dashboard Web（地址为 https://192.168.1.146:31001/），输入 Token 登录即可，如图 3-3 所示。

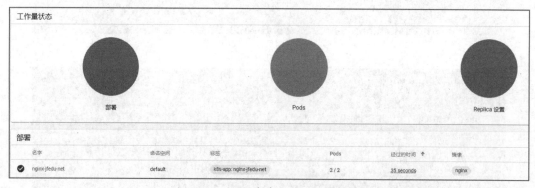

图 3-3　Kubernetes Web 界面展示

3.4 Kubernetes YAML 文件详解

　　Kubernetes YAML 配置文件主要分为基本标签、元数据标签、资源内容 3 部分，要想对 Kubernetes 熟练地掌握，必须要了解 YAML 配置文件中常见的参数和指令的含义。

　　（1）基本标签主要在文件起始位置。例如：

```
apiVersion: v1                              #版本号,例如v1
kind: Namespace                             #类型或者控制器
```

　　（2）元数据标签主要在文件中部位置。例如：

```
metadata:                                   #数据标签
  name: nginx-deployment
  labels:                                   #子标签
    app: nginx                              #业务容器
```

　　（3）资源内容主要在文件末尾位置。例如：

```
spec:                                       #Pod 中容器的详细定义
  containers:                               #容器列表
  - name: nginx                             #容器名称
    image: nginx                            #容器的镜像名称
    imagePullPolicy: [Always | Never | IfNotPresent]
                                            #获取镜像的策略,其中 Always 表示下载镜像;IfnotPresent 表示优
                                            #先使用本地镜像,否则下载镜像;Never 表示仅使用本地镜像
    command: nginx   #容器的启动命令列表,如果不指定,则使用打包时使用的启动命令
    args: -g daemon off                     #容器的启动命令参数列表
    workingDir: /root/                      #容器的工作目录
    volumeMounts:                           #挂载到容器内部的存储卷配置
    - name: nginx    #引用 Pod 定义的共享存储卷的名称,需要用 volumes[]部分定义的
                     #卷名
      mountPath: /usr/share/nginx/html      #存储卷在容器内 mount 的绝对路径
      readOnly: boolean                     #是否为只读模式
    ports:                                  #需要暴露的端口库号列表
    - name: nginx                           #端口号名称
      containerPort: 80                     #容器需要监听的端口号
      hostPort: 80   #容器所在主机需要监听的端口号,默认为与 Container 相同
      protocol: TCP                         #端口协议,支持 TCP 和 UDP,默认为 TCP
    env:                                    #容器运行前需设置的环境变量列表
    - name: WEB                             #环境变量名称
      value: www.jfedu.net                  #环境变量的值
```

```yaml
      resources:              #资源限制和请求的设置
        limits:               #资源限制的设置
          cpu: 1000m          #CPU的限制,单位为core数,将用于docker run --cpu-shares
                              #参数
          memory: 1024m       #内存限制,单位可以为Mib/Gib,将用于docker run -memory
                              #参数
        requests:             #资源请求的设置
          cpu: 100m           #CPU请求,容器启动的初始可用数量
          memory: 1024m       #内存清楚,容器启动的初始可用数量
      livenessProbe:          #对Pod内各容器健康检查的设置,当探测无响应几次后将自动重启
                              #该容器,检查方法有exec、httpGet和tcpSocket,对一个容
                              #器只需设置其中一种方法即可
        exec:                 #将Pod容器内检查方式设置为exec
          command: [string]   #exec方式需要制定的命令或脚本
        httpGet:    #将Pod内各容器健康检查方法设置为HttpGet,需要制定path和port
          path: string
          port: number
          host: string
          scheme: string
          HttpHeaders:
          - name: string
            value: string
        tcpSocket:            #将Pod内各容器健康检查方式设置为tcpSocket方式
          port: number
         initialDelaySeconds: 0    #容器启动完成后首次探测的时间,单位为s
         timeoutSeconds: 0 #对容器健康检查探测等待响应的超时时间,单位为s,默认为1s
         periodSeconds: 0 #对容器监控检查的定期探测时间设置,单位为s,默认为10s一次
         successThreshold: 0
         failureThreshold: 0
         securityContext:
           privileged:false
    restartPolicy: [Always | Never | OnFailure]
                              #Pod的重启策略,其中,Always表示一旦不管以何种方式终止
                              #运行,kubelet都将重启;OnFailure表示只有Pod以非0退
                              #出码退出才重启;Never表示不再重启该Pod
    nodeSelector: obeject     #设置nodeSelector,表示将该Pod调度到包含这个label
                              #的node上,以key: value的格式指定
    imagePullSecrets:         #拉取镜像时使用的secret名称,以key: secretkey格式指定
    - name: string
    hostNetwork:false         #是否使用主机网络模式,默认为false,如果设置为true,表示
                              #使用宿主机网络
```

```
    volumes:                 #在该 Pod 上定义共享存储卷列表
    - name: string           #共享存储卷名称（volumes 类型有很多种）
      emptyDir: {}           #类型为 emtyDir 的存储卷，与 Pod 同生命周期的一个临时目录。
                             #为空值
      hostPath: string       #类型为 hostPath 的存储卷，表示挂载 Pod 所在宿主机的目录
        path: string         #Pod 所在宿主机的目录，将被用于同期中 mount 的目录
      secret:                #类型为 secret 的存储卷，挂载集群与定义的 secret 对象到容器
                             #内部
        secretname: string
        items:
        - key: string
          path: string
      configMap:             #类型为 configMap 的存储卷，挂载预定义的 configMap 对象到
                             #容器内部
        name: string
        items:
        - key: string
          path: string
```

3.5 kubectl 常见指令操作

Kubernetes 云计算平台部署和创建完成后，可以通过 kubectl 指令查看 Pod 和 service 的状态、信息，操作指令如下：

```
#查看 Kubernetes 集群所有的节点信息
kubectl get nodes
#删除 Kubernetes 集群中某个特定节点
kubectl delete nodes/10.0.0.123
#获取 Kubernetes 集群命名空间
kubectl get namespace
#获取 Kubernetes 所有命名空间有哪些部署
kubectl get deployment --all-namespaces
#查看 nginx 部署详细的信息
kubectl describe deployments/nginx -n default
#将 nginx 部署的镜像更新至 nginx1.19 版本
kubectl -n default set image deployments/nginx nginx=nginx:v1.19
#将 nginx 部署的 Pod 组容器副本数调整为 5 个
kubectl patch deployment nginx -p '{"spec":{"replicas":3}}' -n default
#获取 nginx 部署的 yaml 配置，输出到 nginx.yaml 文件
kubectl get deploy nginx -o yaml --export >nginx.yaml
```

```
#修改 nginx.yaml 文件,重新应用现有的 nginx 部署
kubectl apply -f nginx.yaml
#获取所有命名空间的详细信息、VIP、运行时间等
kubectl get svc --all-namespaces
#获取所有 Pod 所属的命名空间
kubectl get pods --all-namespaces
#获取所有命名空间的 Pod 详细 IP 信息
kubectl get pods -o wide --all-namespaces
#查看 dashboard 服务详细信息
kubectl describe service/kubernetes-dashboard --namespace="kube-system"
#获取 dashboard 容器详细信息
kubectl describe pod/kubernetes-dashboard-530803917-816df --namespace="kube-system"
#强制删除 dashboard 容器资源
kubectl delete pod/kubernetes-dashboard-530803917-816df --namespace="kube-system" --grace-period=0 --force
#强制删除一个 node 上的所有容器、服务、部署,不再接受新的 Pod 进程资源创建
kubectl drain 10.0.0.122 --force --ignore-daemonsets --delete-local-data
#恢复 node 上接受新的 Pod 进程资源创建
kubectl uncordon 10.0.0.122
```

3.6 Kubernetes 本地私有仓库实战

Docker 仓库主要用于存放 Docker 镜像。Docker 仓库分为公共仓库和私有仓库,基于 Registry 可以搭建本地私有仓库,使用私有仓库的优点如下:

（1）节省网络带宽,不用去 Docker 官网仓库下载每个镜像。

（2）从本地私有仓库中下载 Docker 镜像。

（3）组件公司内部私有仓库,方便各部门使用,服务器管理更加统一。

（4）可以基于 GIT 或者 SVN、Jenkins 更新本地 Docker 私有仓库镜像版本。

官方提供 Docker Registry 构建本地私有仓库,目前最新版本为 Registry v2。最新版的 Docker 已不再支持 Registry v1。Registry v2 使用 Go 语言编写,在性能和安全性上作了很多优化,重新设计了镜像的存储格式。以下为在 192.168.1.148 服务器上构建 Docker 本地私有仓库的方法及步骤。

（1）下载 Docker Registry 镜像。命令如下:

```
docker pull registry
```

（2）启动私有仓库容器。命令如下：

```
mkdir -p /data/registry/
docker run -itd -p 5000:5000 -v /data/registry:/var/lib/registry
docker.io/registry
```

Docker 私有仓库创建和启动命令如图 3-4 所示。

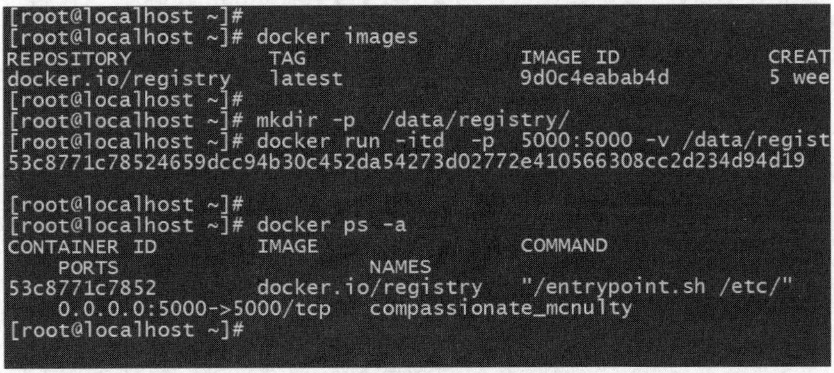

图 3-4 Docker 仓库配置实战

默认情况下，会将仓库存放于容器内的/tmp/registry 目录下，这样如果容器被删除，则存放于容器中的镜像也会丢失，所以一般情况下会指定本地一个目录挂载到容器内的/var/lib/registry 下。

（3）上传镜像至本地私有仓库。

客户端上传镜像至本地私有仓库。下面以 busybox 镜像为例，将 busybox 上传至私有仓库服务器。

```
docker   pull   busybox
docker   tag    busybox   192.168.1.148:5000/busybox
docker   push   192.168.1.148:5000/busybox
```

（4）检测本地私有仓库。

```
curl -XGET http://192.168.1.148:5000/v2/_catalog
curl -XGET http://192.168.1.148:5000/v2/busybox/tags/list
```

（5）在 Kubernetes 集群中其他节点（Docker 客户端）的主机上添加本地仓库地址，修改/etc/docker/daemon.json 文件，添加如下代码，同时重启 Docker 服务即可。

```
{
"insecure-registries":["192.168.1.148:5000"]
}
```

第 4 章 Kubernetes 核心组件 service 实战

4.1 Kubernetes service 概念

service 是 Kubernetes 最核心的概念，通过创建 service，可以为一组具有相同功能的 Pod 应用提供统一的访问入口，并且将请求进行负载分发到后端的各个容器应用上。

在 Kubernetes 中，在受到 RC 调控的时候，Pod 副本是变化的，对于 Pod 容器的 IP 也是变化的，如发生迁移或者伸缩的时候。这对于 Pod 的访问者来说是不可接受的。

Kubernetes 中的 service 是一种抽象概念，它定义了一个 Pod 逻辑集合以及访问它们的策略，service 与 Pod 的关联同样是基于 label 来完成的。service 的目标是提供一种桥梁，它会为访问者提供一个固定访问地址，用于在访问时重定向到相应的后端，这使得非 Kubernetes 原生应用程序在无须为 Kubemces 编写特定代码的前提下，可以轻松访问后端。

service 同 RC 一样，都是通过 label 来关联 Pod 的。当在 service 的 YAML 文件中定义了该 service 的 selector 中的 label 为 app:jfedu-app 时，这个 service 会将 Pod->metadata->labels 中 label 为 app:jfedu-app 的 Pod 作为分发请求的后端。

当 Pod 发生变化（增加、减少、重建等）时，service 会及时更新。这样一来，service 就可以作为 Pod 的访问入口，起到代理服务器的作用，而对于访问者来说，通过 service 进行访问，无须关注后端 Pod 容器是否有变化或更新。

4.2 Kubernetes service 实现方式

Kubernetes 分配给 service 的固定 IP 是一个虚拟 IP，在外部是无法寻址的。在真实的系统实现上，Kubernetes 通过 kube-proxy 组件实现虚拟 IP 路由及转发。所以在之前集群部署的环节，在每个 node 上均部署了 Proxy 这个组件，从而实现了 Kubernetes 层级的虚拟转发网络。

Kubernetes 为每个 service 分配一个唯一的 ClusterIP，所以当使用 ClusterIP: port 的组合访问一个 service 时，不管 port 是什么，这个组合是不可能发生重复的。另一方面，kube-proxy 为每个 service 真正打开的是一个绝对不会重复的随机端口，用户在 service 描述文件中指定的访问端口会被映射到这个随机端口上。这就是为什么用户可以在创建 service 时随意指定访问端口。

在 Kubernetes 集群中，Pod 的 IP 是在 docker0 网段动态分配的，当进行重启、扩容等操作时，IP 地址会随之变化。当某个 Pod（frontend）需要访问其依赖的另外一组 Pod（backend）时，如果 backend 的 IP 发生变化，如何保证 frontend 到 backend 的正常通信变得非常重要，此时需要借助 service 实现统一访问。

service 的 Virtual IP 是由 Kubernetes 虚拟出来的内部网络，外部是无法寻址的。但是有些服务又需要被外部访问，如 Web 前段，这时就需要加一层网络转发，即外网到内网的转发。Kubernetes 提供了 ClusterIP、nodePort、LoadBalancer、Ingress 四种方式。

（1）ClusterIP：在 Kubernetes 平台创建容器时，选择内部服务，默认会创建一个 service，其 IP 类型为 ClusterIP，该 IP（virtual IP Address，虚拟 IP 地址）只能在 Kubernetes 集群内部使用。

（2）NodePort：Kubernetes 会在每个 node 上暴露出一个端口：NodePort，外部网络可以通过（任一 node）[NodeIP]:[NodePort]访问到后端的 service。

（3）LoadBalancer：在 NodePort 基础上，Kubernetes 可以请求底层云平台创建一个负载均衡器，将每个 node 作为后端进行服务分发。该模式需要底层云平台（如 GCE）支持。

（4）Ingress：是一种 HTTP 方式的路由转发机制，由 Ingress Controller 和 HTTP 代理服务器组合而成。Ingress Controller 实时监控 Kubernetes API，实时更新 HTTP 代理服务器的转发规则。HTTP 代理服务器有 GCE Load-Balancer、HAProxy、Nginx 等开源方案。

4.3 service 实战：ClusterIP 案例演练

ClusterIP 模式通常又称为内部服务，当然外部服务方式也有 ClusterIP。默认创建一个 service 内部服务，Kubernetes 将会在集群中生成一个 VIP 地址，该 VIP 是不能寻址的，只能集群内部访问，其原理是通过 kube-proxy 调用 iptables 防火墙规则添加 NAT 映射，此时用户在 Kubernetes 内部通过 VIP 地址就可以访问到 service 均衡后端的 Pod 容器服务。

ClusterIP 模式在每个 node 节点上不会配置 IP 地址，同时端口也不会显示，只能通过 Web 界面或者其他命令行指令查看。

（1）ClusterIP 内部服务案例演练配置实战如图 4-1 所示。选择内部网络，操作界面如下：

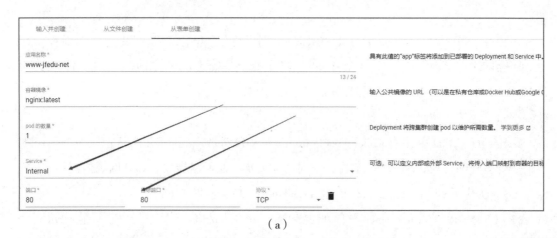

(a)

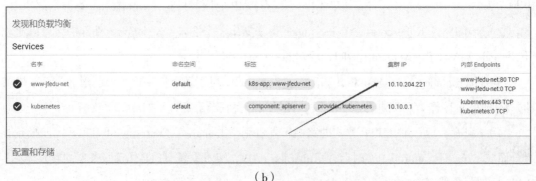

(b)

图 4-1　Kubernetes 创建容器设置为内部服务

（2）在 Kubernetes 任意节点上通过 Cluster IP+80 端口访问，如图 4-2 所示。

```
[root@node1 ~]# curl 10.10.204.221
<!DOCTYPE html>
<html>
<head>
<title>Welcome to nginx!</title>
<style>
    body {
        width: 35em;
        margin: 0 auto;
        font-family: Tahoma, Verdana, Arial, sans-serif;
    }
</style>
</head>
<body>
```

图 4-2　Kubernetes 内部服务 VIP 访问效果

4.4　service 实战：NodePort 案例演练

NodePort 模式下，Kubernetes 将会在每个 node 上打开一个端口，且每个 node 的端口都是一样的，通过 <NodeIP>:NodePort 的方式，Kubernetes 集群外部的程序可以访问 service。

NodePort 中每个 node 节点的端口有很多（0~65535 个），Kubernetes 外部服务创建如图 4-3 所示。

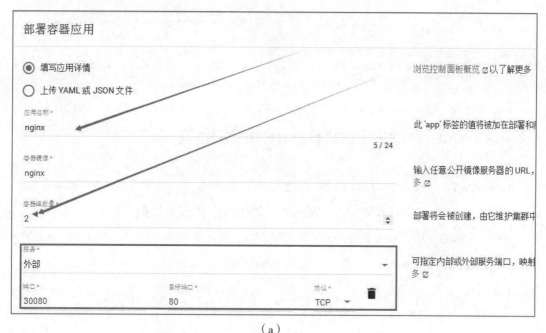

（a）

图 4-3　Kubernetes 外部服务创建

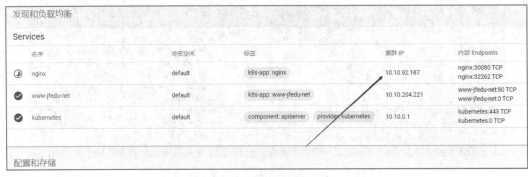

（b）

图 4-3 （续）

在 Kubernetes 任意节点上通过 node IP+32262 端口访问，如图 4-4 所示。

图 4-4 Kubernetes 外部服务访问和验证

4.5 service 实战：LoadBalancer 案例演练

LoadBalancer service 是 Kubernetes 深度结合云平台的一个组件；当使用 LoadBalancer service 暴露服务时，实际上是通过向底层云平台申请创建一个负载均衡器来向外暴露服务。

目前 LoadBalancer service 支持的云平台已经相对完善，如国外的 GCE、DigitalOcean，国内的阿里云，私有云 Openstack 等。由于 LoadBalancer service 深度结合了云平台，所以能在一些云平台上使用。

LoadBalancer 会分配 ClusterIP 和 NodePort，通过 Cloud Provider 实现 LoadBalancer 设备的配制，并且在 LoadBalancer 设备配置中将<NodeIP>:NodePort 作为 Pool Member，LoadBalancer 设备依据转发规则将流量转到节点的 NodePort，如图 4-5 所示。

创建 LoadBalancer service，操作指令如下，如图 4-5 所示。

```
kubectl expose deployment nginxv1 --port=8081 --target-port=80
--type=LoadBalancer
```

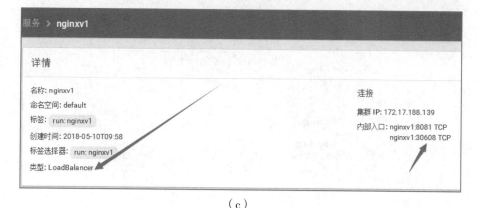

图 4-5　Kubernetes 创建 LoadBalancer service

4.6 service 实战：Ingress 案例演练

Ingress 是 Kubernetes 1.2 后才出现的，通过 Ingress 用户可以使用 Nginx 等开源的反向代理服务器实现对外暴露服务，后面 Traefik 用的也是 Ingress。使用 Ingress 通常需要以下 3 个组件。

（1）Ingress Controller：可以将 Ingress Controller 视作监视器，Ingress Controller 通过不断地与 Kubernetes API 交互，实时感知后端 service 和 Pod 的变化。例如，新增和减少 Pod, service 增加与减少等。当得到这些变化信息后，Ingress Controller 会跟 Ingress 生成相应的配置，然后将生成的配置刷新到反向代理服务器（Nginx）配置文件中，达到服务自动发现的作用。

（2）Ingress：Ingress 主要用来实现规则定义。例如，某个域名对应某个 service，即当某个域名的请求进来时转发给某个 service，这个规则将与 Ingress Controller 结合，然后 Ingress Controller 将其动态写入反向代理负载均衡器配置中，从而实现整体的服务发现和负载均衡。

（3）反向代理服务器（Nginx）：反向代理服务器种类很多，可以采用 Nginx、HAProxy、Apache 等，通常 Nginx 使用非常多。Nginx 的特点有轻量级、高性能、配置简单、管理便捷等，此处不需要另外部署一套 Nginx，因为在部署 Ingress Controller 服务时，会自动部署 Nginx。

用户访问某个网站域名时，请求首先会到达反向代理服务器，Ingress Controller 通过与 Ingress 交互得知某个域名对应哪个 service，再通过与 Kubernetes API 交互得知 service 地址等信息。综合以后，生成配置文件并实时写入反向代理服务器，然后反向代理服务器重新载入该规则便可实现服务发现，即动态映射，如图 4-6 所示。

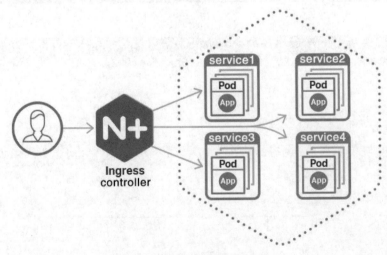

图 4-6　Nginx Ingress 内部结构图

Ingress Controller 通过和 Kubernetes API 交互，动态感知集群中 Ingress 规则变化，然后读取它，按照自定义的规则，规则就是写明了哪个域名对应哪个 service，生成一段 Nginx 配置。

再写到 Nginx-ingress-controller 的 Pod 中，这个 Ingress Controller 的 Pod 中运行着一个 Nginx 服务，控制器会把生成的 Nginx 配置写入/etc/nginx.conf 文件，然后重新载入使配置生效。以此达到域名分别配置和动态更新的目的。

无论如何请求，反向代理服务器的对外端口最终会暴露在固定的 node 上，同时以 Hostport 方式监听 80 端口，就解决了其他方式部署不确定反向代理服务器在哪儿的问题，同时访问每个 node 的 80 端口都能正确解析请求，如果前端再放置 Nginx，就又实现了一层负载均衡。

（1）部署 Ingress-nginx-controller 控制服务，从官网下载 YAML 配置文件，操作指令如下：

```
wget https://raw.githubusercontent.com/kubernetes/ingress-nginx/nginx-0.30.0/deploy/static/mandatory.yaml
```

（2）修改文件 mandatory.yaml，在 213 行中加入如下代码：

```
hostNetwork: true    #在 Pod 中使用 hostNetwork:true 配置网络,Pod 中运行的应用程序
                     #可以直接看到宿主主机的网络接口,宿主主机所在的局域网上所有网络
                     #接口都可以访问该应用程序
```

（3）创建并应用 YAML 文件，操作指令如下：

```
kubectl apply -f mandatory.yaml
kubectl get pod -n ingress-nginx
```

（4）将 Ingress-nginx-controller 暴露为一个 service 资源对象，在命令行窗口打开文件 service-nodeport.yaml，YAML 代码如下：

```
apiVersion: v1
kind: Service
metadata:
  name: ingress-nginx
  namespace: ingress-nginx
  labels:
    app.kubernetes.io/name: ingress-nginx
    app.kubernetes.io/part-of: ingress-nginx
spec:
  type: NodePort
  ports:
  - name: http
    port: 80
    targetPort: 80
```

```
    protocol: TCP
  - name: https
    port: 443
    targetPort: 443
    protocol: TCP
  selector:
    app.kubernetes.io/name: ingress-nginx
    app.kubernetes.io/part-of: ingress-nginx
```

(5)执行如上 Ingress-nginx Service yaml 文件,操作指令如下:

```
kubectl apply -f service-nodeport.yaml
kubectl get svc -n ingress-nginx
```

(6)创建 ingress 规则 YAML 文件,关联 v1-jfedu-net 和 v2-jfedu-net 服务即可。ingress.yaml 配置文件代码如下:

```
apiVersion: extensions/v1beta1
kind: Ingress
metadata:
  name: v1-jfedu-net
spec:
  rules:
  - host: v1.jfedu.net
    http:
      paths:
      - path: /
        backend:
          serviceName: v1-jfedu-net
          servicePort: 80
---
apiVersion: extensions/v1beta1
kind: Ingress
metadata:
  name: v2-jfedu-net
spec:
  rules:
  - host: v2.jfedu.net
    http:
      paths:
      - path: /
        backend:
          serviceName: v2-jfedu-net
          servicePort: 80
```

（7）创建 Ingress 关联 v1-jfedu-net 和 v2-jfedu-net 服务，如图 4-7 所示，操作指令如下：

```
kubectl apply -f ingress.yaml
kubectl get ingresses
```

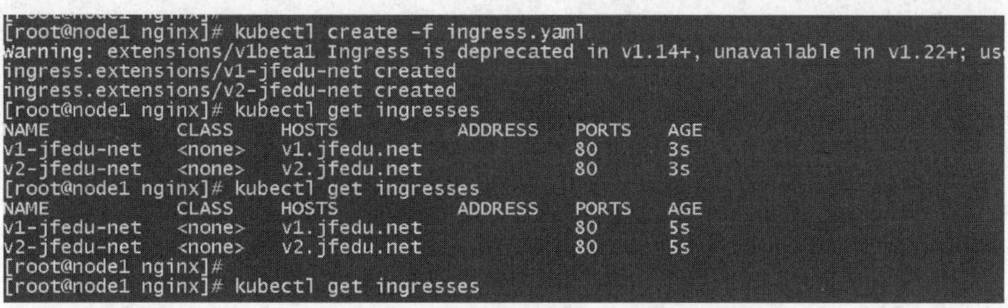

图 4-7 Kubernetes Nginx Ingress 配置实战

（8）分别创建两个应用部署和容器，并设置服务为内部服务，部署名称分别为 v1-jfedu-net 和 v2-jfedu-net，如图 4-8 所示。

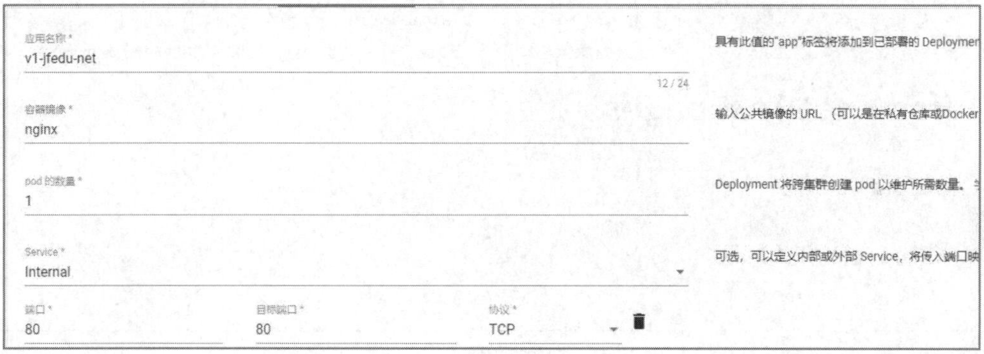

(a)

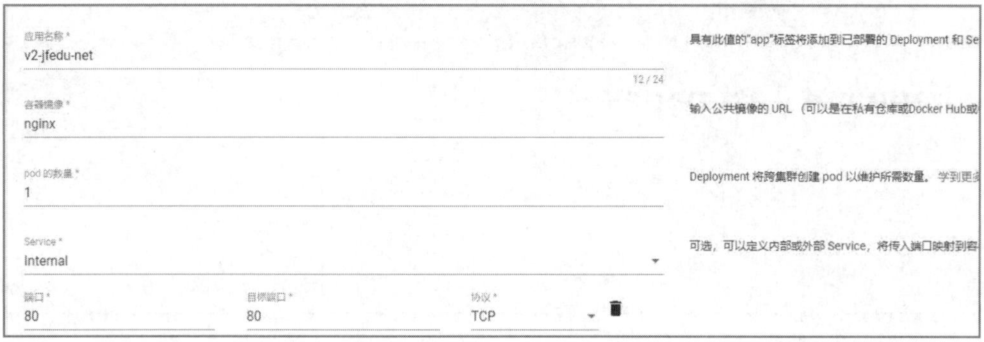

(b)

图 4-8 Kubernetes 创建 v2-jfedu-net 应用和服务

（9）查看 Ingress server 服务对外监听的 node IP 端口为 80，在客户端 hosts 文件中添加 v1.jfedu.net、v2.jfedu.net 域名映射，通过浏览器访问，如图 4-9 所示。

（a）

（b）

（c）

图 4-9　Kubernetes Nginx Ingress 案例实战

4.7 Kubernetes Traefik 案例实战

由于微服务架构以及 Docker 技术和 Kubernetes 编排工具最近几年才开始逐渐流行，所以一开始的反向代理服务器（如 Nginx、Apache）并未对其提供支持，所以才会出现 Ingress Controller 来作为 Kubernetes 和前端反向代理服务器（如 Nginx）之间的衔接。

即 Ingress Controller 的存在就是为了既能与 Kubernetes 交互，又能写 Nginx 配置，还能重新载入配置，这是一种折中方案。而最近开始出现的 Traefik 就提供了对 Kubernetes 的支持，也就是说，Traefik 本身就能与 Kubernetes API 交互，感知后端变化，如果使用 Traefik，Ingress Controller 就没有用了，所以，Kubernetes Traefik service 案例实战如图 4-10 所示。

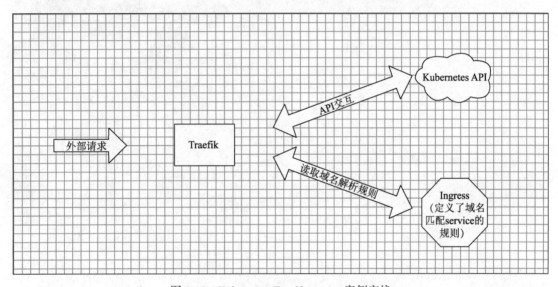

图 4-10　Kubernetes Traefik service 案例实战

Kubernetes Traefik 在企业生产环境中也被广泛采用，部署 Traefik service 也非常简单，操作的方法和步骤如下。

（1）创建 Traefik rbac 验证配置，rbac.yaml 文件内容如下：

```
---
kind: ClusterRole
apiVersion: rbac.authorization.Kubernetes.io/v1beta1
metadata:
  name: traefik-ingress-controller
```

```yaml
    rules:
      - apiGroups:
          - ""
        resources:
          - services
          - endpoints
          - secrets
        verbs:
          - get
          - list
          - watch
      - apiGroups:
          - extensions
        resources:
          - ingresses
        verbs:
          - get
          - list
          - watch
      - apiGroups:
          - extensions
        resources:
          - ingresses/status
        verbs:
          - update
---
kind: ClusterRoleBinding
apiVersion: rbac.authorization.Kubernetes.io/v1beta1
metadata:
  name: traefik-ingress-controller
roleRef:
  apiGroup: rbac.authorization.Kubernetes.io
  kind: ClusterRole
  name: traefik-ingress-controller
subjects:
- kind: ServiceAccount
  name: traefik-ingress-controller
  namespace: kube-system
```

（2）以 DaemonSet 的方式在每个 node 上启动一个 Traefik，并使用 hostPort 的方式让其监听每个 node 的 80 端口。创建部署 YAML 文件 traefik-depolyment.yaml，代码如下：

```yaml
---
apiVersion: v1
kind: ServiceAccount
metadata:
  name: traefik-ingress-controller
  namespace: kube-system
---
kind: DaemonSet
apiVersion: apps/v1
metadata:
  name: traefik-ingress-controller
  namespace: kube-system
  labels:
    Kubernetes-app: traefik-ingress-lb
spec:
  selector:
    matchLabels:
      Kubernetes-app: traefik-ingress-lb
      name: traefik-ingress-lb
  template:
    metadata:
      labels:
        Kubernetes-app: traefik-ingress-lb
        name: traefik-ingress-lb
    spec:
      serviceAccountName: traefik-ingress-controller
      terminationGracePeriodSeconds: 60
      containers:
      - image: traefik:v1.7
        name: traefik-ingress-lb
        ports:
        - name: http
          containerPort: 80
          hostPort: 80
        - name: admin
```

```
              containerPort: 8080
              hostPort: 8080
          securityContext:
            capabilities:
              drop:
              - ALL
              add:
              - NET_BIND_SERVICE
          args:
          - --api
          - --kubernetes
          - --logLevel=INFO
---
kind: Service
apiVersion: v1
metadata:
  name: traefik-ingress-service
  namespace: kube-system
spec:
  selector:
    Kubernetes-app: traefik-ingress-lb
  ports:
    - protocol: TCP
      port: 80
      name: web
    - protocol: TCP
      port: 8080
      name: admin
```

（3）以 Deployment 的方式启动一个 Traefik，并使用 hostPort 的方式让其监听每个 node 的 80 端口。创建部署 YAML 文件 traefik-depolyment.yaml，代码如下：

```
---
apiVersion: v1
kind: ServiceAccount
metadata:
  name: traefik-ingress-controller
  namespace: kube-system
---
```

```yaml
kind: Deployment
apiVersion: apps/v1
metadata:
  name: traefik-ingress-controller
  namespace: kube-system
  labels:
    Kubernetes-app: traefik-ingress-lb
spec:
  replicas: 1
  selector:
    matchLabels:
      Kubernetes-app: traefik-ingress-lb
  template:
    metadata:
      labels:
        Kubernetes-app: traefik-ingress-lb
        name: traefik-ingress-lb
    spec:
      serviceAccountName: traefik-ingress-controller
      terminationGracePeriodSeconds: 60
      containers:
      - image: traefik:v1.7
        name: traefik-ingress-lb
        ports:
        - name: http
          containerPort: 80
        - name: admin
          containerPort: 8080
        args:
        - --api
        - --kubernetes
        - --logLevel=INFO
---
kind: Service
apiVersion: v1
metadata:
  name: traefik-ingress-service
  namespace: kube-system
```

```
spec:
  type: NodePort
  selector:
    Kubernetes-app: traefik-ingress-lb
  ports:
    - protocol: TCP
      port: 80
      name: web
      targetPort: 80
    - protocol: TCP
      port: 8080
      name: admin
```

其中 Traefik 监听 node 的 80 和 8080 端口，80 提供正常服务，8080 是其自带的用户界面。

（4）部署 Ingress 规则，Ingress Controller 是无须部署的，所以直接部署 Ingress，编写 traefik-ingress.yaml 文件，文件代码如下：

```
apiVersion: extensions/v1beta1
kind: Ingress
metadata:
  name: v1-jfedu-net
spec:
  rules:
    - host: v1.jfedu.net
      http:
        paths:
          - path: /
            backend:
              serviceName: v1-jfedu-net
              servicePort: 80
---
apiVersion: extensions/v1beta1
kind: Ingress
metadata:
  name: v2-jfedu-net
spec:
  rules:
    - host: v2.jfedu.net
      http:
```

```
      paths:
      - path: /
        backend:
          serviceName: v2-jfedu-net
          servicePort: 80
```

实际上，因为集群中已经存在了相应的名为 v1-jfedu-net 和 v2-jfedu-net 的 service，对应的 service 后端也有很多 Pod，所以这里就不再具体介绍部署实际业务容器（v1-jfedu-net、v2-jfedu-net）的过程了，测试的时候，只需要把这两个 service 替换成自己业务的 service 即可。

（5）部署 Traefik UI。Traefik 本身还提供了一套用户界面，同样以 Ingress 方式暴露，只需创建一下 traefik-ui.yaml 文件。traefik-ui.yaml 文件内容如下：

```
---
apiVersion: v1
kind: Service
metadata:
  name: traefik-web-ui
  namespace: kube-system
spec:
  selector:
    Kubernetes-app: traefik-ingress-lb
  ports:
  - name: web
    port: 80
    targetPort: 8080
---
apiVersion: extensions/v1beta1
kind: Ingress
metadata:
  name: traefik-web-ui
  namespace: kube-system
spec:
  rules:
  - host: traefik-ui.minikube
    http:
      paths:
      - path: /
        backend:
```

```
            serviceName: traefik-web-ui
            servicePort: web
```

（6）创建两个应用部署和容器，并且设置服务为内部服务，部署名称分别为 v1-jfedu-net 和 v2-jfedu-net，操作如图 4-11 所示。

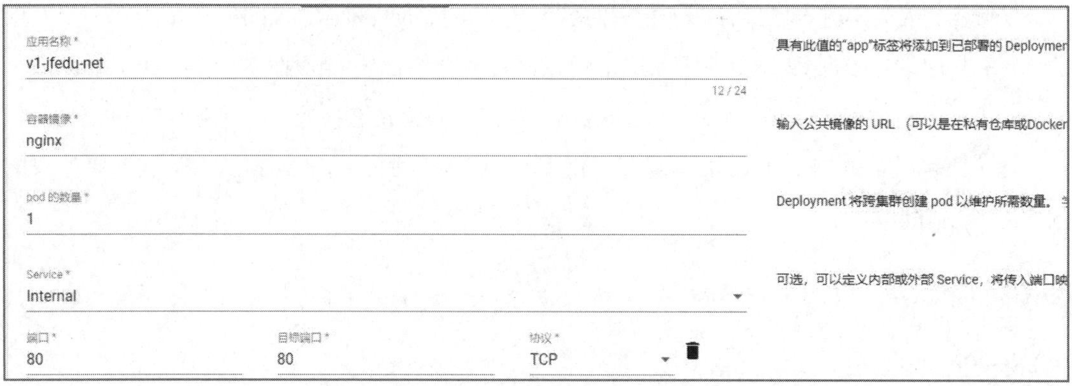

（a）Kubernetes 创建内部服务 v1-jfedu-net

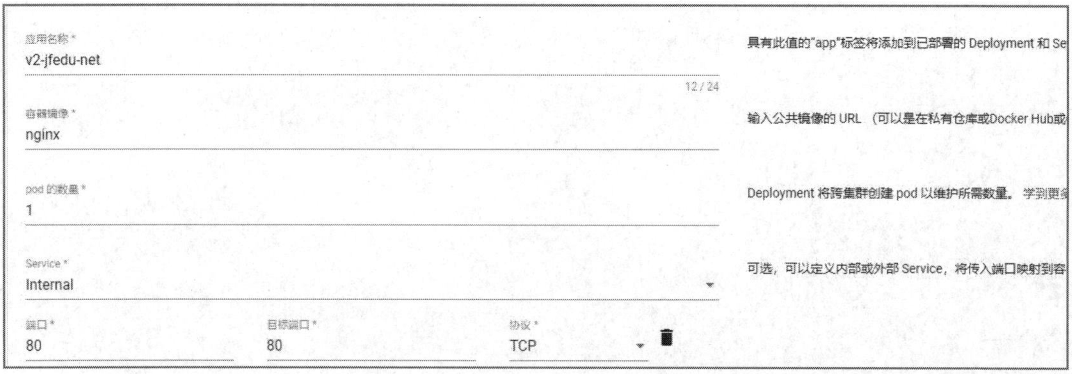

（b）Kubernetes 创建内部服务 v2-jfedu-net

图 4-11　创建两个内部服务

（7）经过以上操作，Kubernetes Traefik service 所需配置文件均创建完成，执行以下指令使其生效即可，操作指令如下：

```
kubectl apply -f rbac.yaml
kubectl apply -f traefik-depolyment.yaml
kubectl apply -f traefik-ingress.yaml
kubectl apply -f traefik-ui.yaml
```

（8）查看 Traefik Web 服务对外监听的 node IP 端口为 32405（需要修改），在客户端 hosts

文件中添加 v1.jfedu.net、v2.jfedu.net 域名映射，通过浏览器访问，如图 4-12 所示。

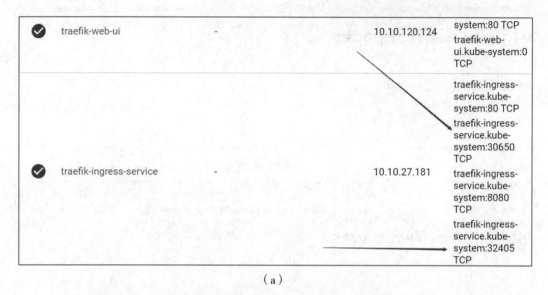

（a）

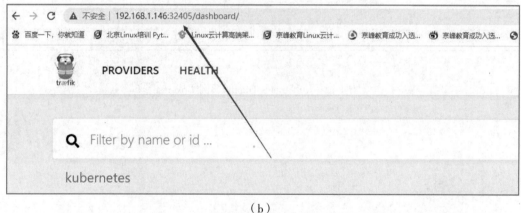

（b）

图 4-12　Kubernetes Traefik 案例实战 1

（9）查看以上 service，NodePort 端口为 30650，即访问虚拟主机需要带上该端口。如何将其端口修改为 80 端口呢？修改 master 节点上的 API server 配置文件/etc/kubernetes/manifests/kube-apiserver.yaml，在 command 字段 kube-apiserver 下添加如下代码：

```
- --service-node-port-range=1-65535
```

（10）分别设置两个部署应用的网站内容，在客户端添加 hosts 绑定到任意一个 node 节点 IP，浏览器访问两个域名，如图 4-13 所示。

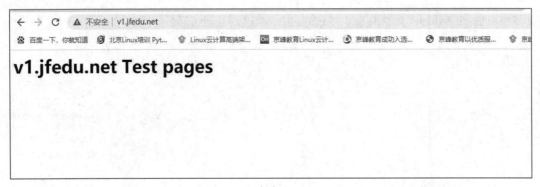

(a)

(b)

图 4-13　Kubernetes Traefik 案例实战 2

第 5 章 Kubernetes 容器升级实战

5.1 Kubernetes 容器升级概念

传统的网站升级更新，通常是将服务全部下线，业务停止后再更新版本和配置，然后重新启动并提供服务。这样的模式已经完全不能满足发展需求了。

高并发、高可用系统普及的今天，服务的升级更新至少要做到"业务不中断"。而滚动更新（Rolling-update）恰好是满足这一需求的系统更新升级方案。

滚动更新就是针对多实例服务的一种不中断服务的更新升级方式。一般情况下，对于多实例服务，滚动更新采用对各个实例逐个进行单独更新，而非同一时刻对所有实例进行全部更新的方式。

对于 Kubernetes 集群部署的 service 来说，Rolling-update 就是指一次仅更新一个 Pod，然后逐个进行更新，而不是在同一时刻将该 service 下面的所有 Pod 关闭，然后更新。逐个更新可以避免业务中断。

Kubernetes 在 kubectl cli 工具中仅提供了对 Replication controller 的 Rolling-update 支持，通过命令 kubectl -help 查看指令信息，如图 5-1 所示。

5.2 Kubernetes 容器升级实现方式

（1）查看部署列表，获取部署应用名称。

```
kubectl get deployments -n default
```

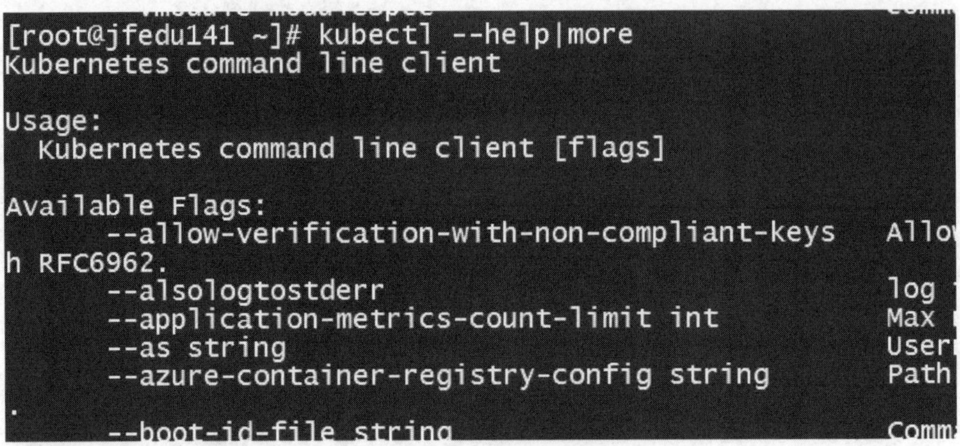

```
[root@jfedu141 ~]# kubectl --help|more
Kubernetes command line client

Usage:
  Kubernetes command line client [flags]

Available Flags:
      --allow-verification-with-non-compliant-keys   Allo
h RFC6962.
      --alsologtostderr                              log
      --application-metrics-count-limit int          Max
      --as string                                    User
      --azure-container-registry-config string       Path
      --boot-id-file string                          Comm
```

图 5-1 kubectl 指令帮助信息

（2）查看正在运行的 Pod。

```
kubectl get pods -n default
```

（3）通过 Pod 描述，查看部署程序的当前映像版本。

```
kubectl describe pods -n default
```

（4）仓库源中提前制作最新更新的镜像，执行如下指令，升级镜像版本即可。升级之前一定要保证仓库源中有最新提交的镜像，代码如下，如图 5-2 所示。

```
kubectl -n default set image deployments/nginx-v1 nginx-v1=docker.io/nginx:v1
```

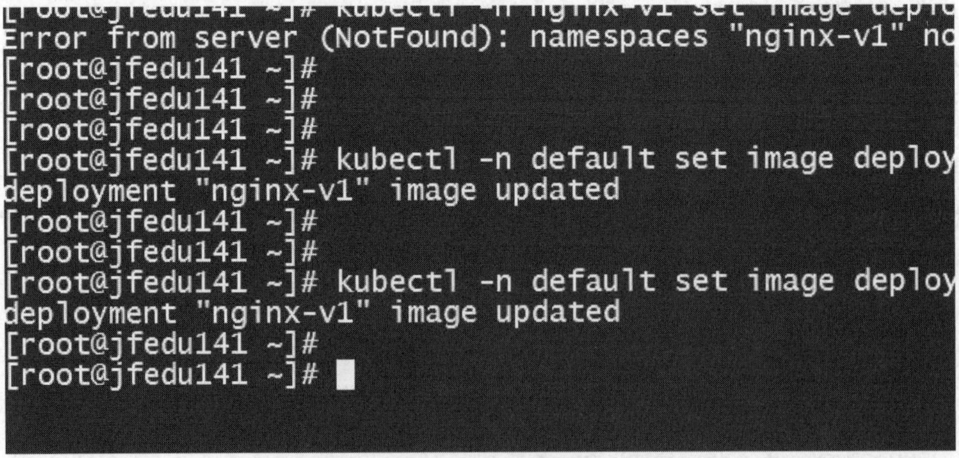

```
[root@jfedu141 ~]# kubectl -n nginx-v1 set image depl
Error from server (NotFound): namespaces "nginx-v1" no
[root@jfedu141 ~]#
[root@jfedu141 ~]#
[root@jfedu141 ~]#
[root@jfedu141 ~]# kubectl -n default set image deploy
deployment "nginx-v1" image updated
[root@jfedu141 ~]#
[root@jfedu141 ~]#
[root@jfedu141 ~]# kubectl -n default set image deploy
deployment "nginx-v1" image updated
[root@jfedu141 ~]#
[root@jfedu141 ~]#
```

图 5-2 kubectl 更新部署应用镜像

5.3 Kubernetes 容器升级测试

（1）创建 nginx-v1 部署，查看 nginx-v1 部署的容器列表，如图 5-3 所示。

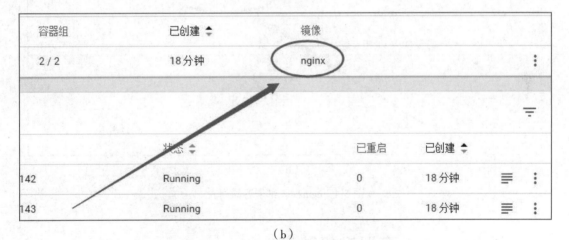

图 5-3　nginx-v1 部署容器列表

（2）镜像更新完成之后，查看 nginx-v1 部署的容器列表，如图 5-4 所示。

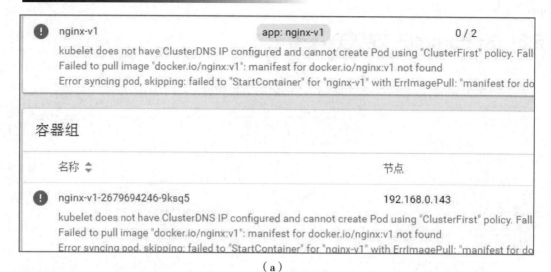

图 5-4　nginx-v1 部署容器列表

5.4　Kubernetes 容器升级验证

（1）检查 Kubernetes 更新 rollout 状态，操作指令如下：

```
kubectl -n default rollout status deployments/nginx-v1
```

（2）检查 Kubernetes Pod 详情，操作指令如下：

```
kubectl describe pods -n default
```

5.5　Kubernetes 容器升级回滚

（1）Kubernetes 部署镜像回滚，操作指令如下：

```
kubectl -n default rollout undo deployments/nginx-v1
```

（2）查看已经部署的版本，操作指令如下：

```
kubectl rollout history deploy/nginx-v1
```

（3）查看某个版本详细信息，如图 5-5 所示，操作指令如下：

```
kubectl rollout history deployment/nginx-v1 --revision=8
```

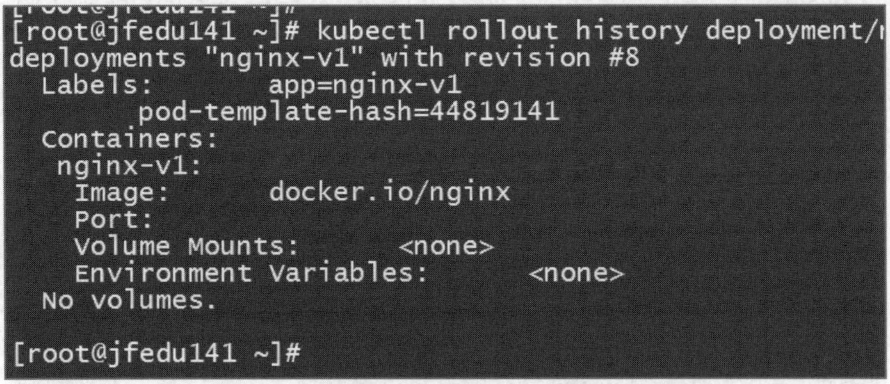

图 5-5　Kubernetes 滚动升级查看版本详细信息

（4）Kubernetes 完美支持回滚至某个版本，还可以通过资源文件进行配置保留的历史版次量。Kubernetes 回滚某个版本，如图 5-6 所示，操作指令如下：

```
kubectl -n default rollout undo deployment/nginx-v1 --to-revision=8
```

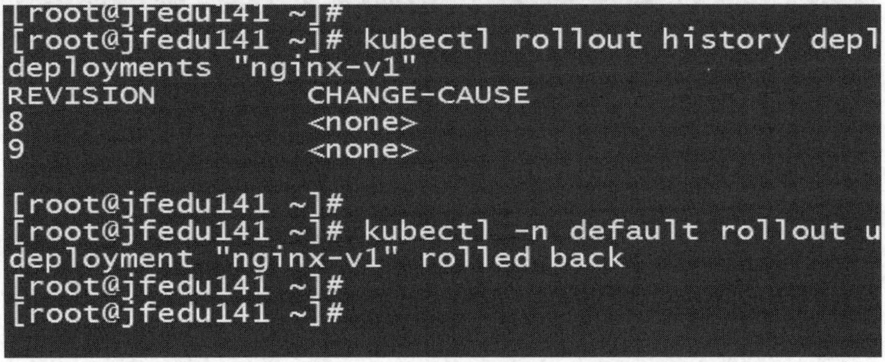

图 5-6　Kubernetes 镜像滚动至指定版本

5.6　Kubernetes 滚动升级和回滚原理

Kubernetes 精确地控制着整个发布过程，分批次有序地进行滚动更新，直到把所有旧的副本全部更新到新版本。实际上 Kubernetes 是通过两个参数精确控制每次滚动的 Pod 数量。

（1）maxSurge：滚动更新过程中运行操作期望副本数的最大 Pod 数量，可以为绝对数值（eg：5），也可以为百分数（eg：10%），但不能为 0，默认为 25%。

（2）maxUnavailable：滚动更新过程中不可用的最大 Pod 数量，可以为绝对数值（eg：5），也可以为百分数（eg：10%），但不能为 0，默认为 25%。

如果未指定这两个可选参数，则 Kubernetes 会使用默认配置，查找默认配置指令，如图 5-7 所示，操作指令如下：

```
kubectl -n default get deployment nginx-v1 -o yaml
```

```
[root@jfedu141 ~]# kubectl -n default get deployment
apiVersion: extensions/v1beta1
kind: Deployment
metadata:
  annotations:
    deployment.kubernetes.io/revision: "10"
  creationTimestamp: 2018-08-22T10:06:43Z
  generation: 22
  labels:
    app: nginx-v1
  name: nginx-v1
  namespace: default
  resourceVersion: "15313"
```

(a)

```
selector:
  matchLabels:
    app: nginx-v1
strategy:
  rollingUpdate:
    maxSurge: 1
    maxUnavailable: 1
  type: RollingUpdate
template:
  metadata:
    creationTimestamp: null
    labels:
      app: nginx-v1
```

(b)

图 5-7　Kubernetes nginx-v1 部署 YAML 文件

（1）查看 Kubernetes nginx-v1 应用部署的概况，如图 5-8 所示。

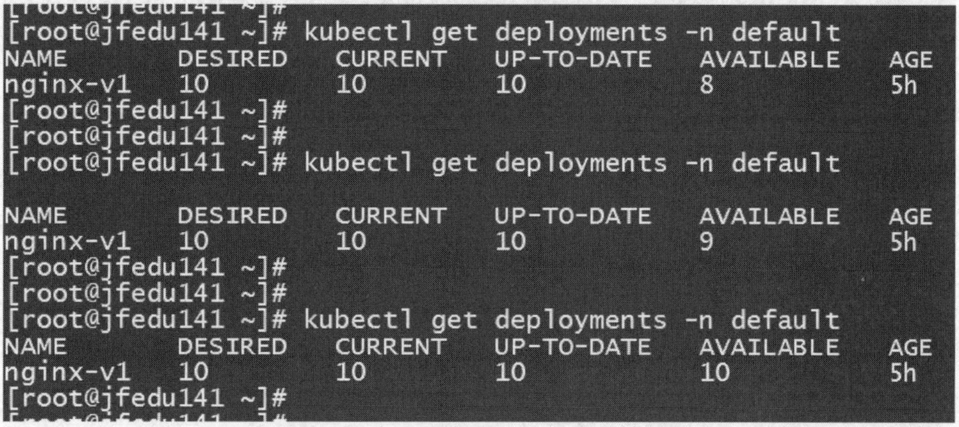

```
[root@jfedu141 ~]# kubectl get deployments -n default
NAME        DESIRED   CURRENT   UP-TO-DATE   AVAILABLE   AGE
nginx-v1    10        10        10           8           5h
[root@jfedu141 ~]#
[root@jfedu141 ~]#
[root@jfedu141 ~]# kubectl get deployments -n default
NAME        DESIRED   CURRENT   UP-TO-DATE   AVAILABLE   AGE
nginx-v1    10        10        10           9           5h
[root@jfedu141 ~]#
[root@jfedu141 ~]#
[root@jfedu141 ~]# kubectl get deployments -n default
NAME        DESIRED   CURRENT   UP-TO-DATE   AVAILABLE   AGE
nginx-v1    10        10        10           10          5h
[root@jfedu141 ~]#
```

(a)

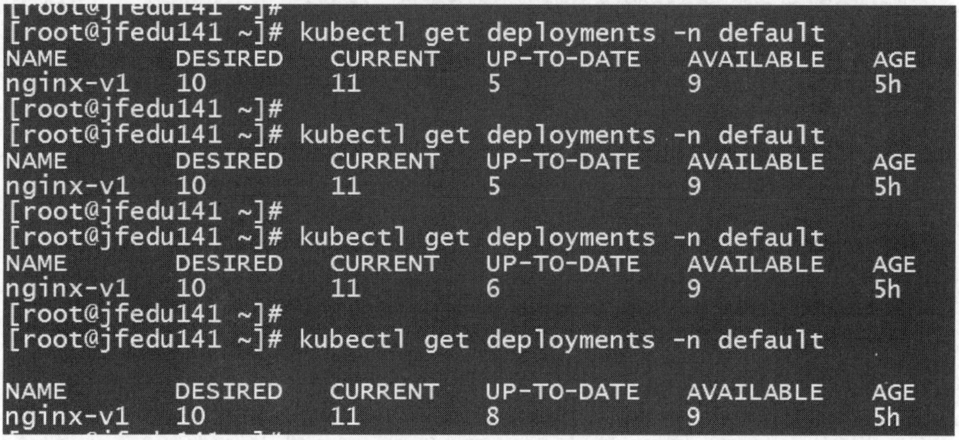

```
[root@jfedu141 ~]#
[root@jfedu141 ~]# kubectl get deployments -n default
NAME        DESIRED   CURRENT   UP-TO-DATE   AVAILABLE   AGE
nginx-v1    10        11        5            9           5h
[root@jfedu141 ~]#
[root@jfedu141 ~]# kubectl get deployments -n default
NAME        DESIRED   CURRENT   UP-TO-DATE   AVAILABLE   AGE
nginx-v1    10        11        5            9           5h
[root@jfedu141 ~]#
[root@jfedu141 ~]# kubectl get deployments -n default
NAME        DESIRED   CURRENT   UP-TO-DATE   AVAILABLE   AGE
nginx-v1    10        11        6            9           5h
[root@jfedu141 ~]#
[root@jfedu141 ~]# kubectl get deployments -n default
NAME        DESIRED   CURRENT   UP-TO-DATE   AVAILABLE   AGE
nginx-v1    10        11        8            9           5h
```

(b)

图 5-8　Kubernetes nginx-v1 应用滚动升级过程

其中，部分参数含义如下。

① DESIRED：最终期望处于 READY 状态的副本数。

② CURRENT：当前的副本总数。

③ UP-TO-DATE：当前完成更新的副本数。

④ AVAILABLE：当前可用的副本数。

（2）查看 Kubernetes nginx-v1 应用镜像更新的效果，如图 5-9 所示，操作指令如下：

```
kubectl -n default describe deployment nginx-v1
```

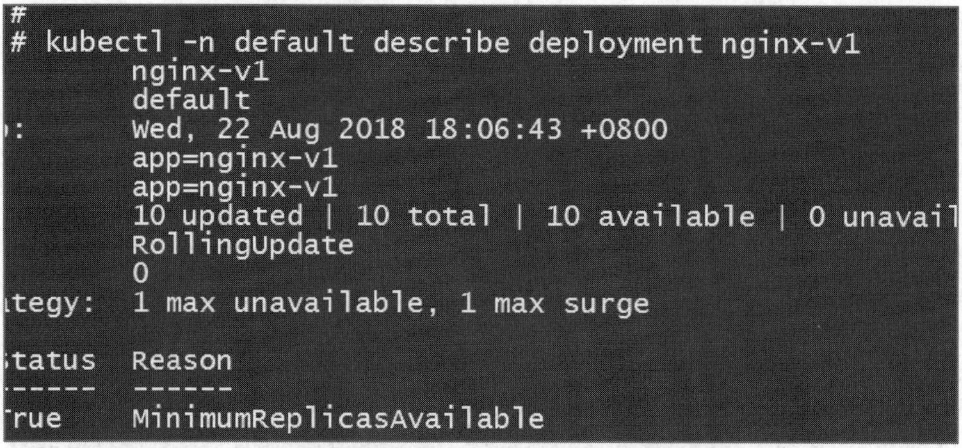

```
# kubectl -n default describe deployment nginx-v1
        nginx-v1
        default
      Wed, 22 Aug 2018 18:06:43 +0800
        app=nginx-v1
        app=nginx-v1
        10 updated | 10 total | 10 available | 0 unavail
        RollingUpdate
        0
tegy:   1 max unavailable, 1 max surge

tatus   Reason
------  ------
rue     MinimumReplicasAvailable
```

图 5-9 Kubernetes nginx-v1 应用镜像更新效果

（3）整个滚动过程是通过控制两个副本集完成的，分别是新副本和旧副本，如图 5-10 所示。名称如下。

① 新副本集：nginx-v1-224846633。

② 旧副本集：nginx-v1-44819141。

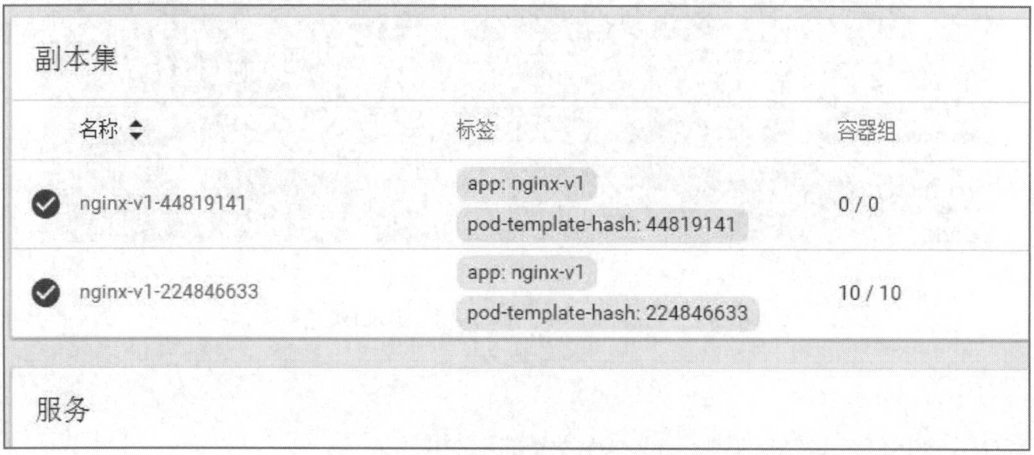

副本集		
名称 ♦	标签	容器组
✓ nginx-v1-44819141	app: nginx-v1 pod-template-hash: 44819141	0 / 0
✓ nginx-v1-224846633	app: nginx-v1 pod-template-hash: 224846633	10 / 10
服务		

图 5-10 Kubernetes nginx-v1 应用副本查看

（4）理想状态下的 Kubernetes 镜像更新滚动的过程如下：

① 创建 1 个新的副本集，并为其分配 3 个新版本的 Pod，使副本总数达到 11，一切正常。

② 通知旧副本集，销毁 2 个旧版本的 Pod，使可用副本总数保持到 8，一切正常。

③ 当 2 个副本销毁成功后，通知新副本集，再新增 2 个新版本的 Pod，使副本总数达到 11，

一切正常。

④ 只要销毁成功，新副本集就会创造新的 Pod，一直循环，直到旧的副本集 Pod 数量为 0。

（5）滚动升级一个服务，实际是创建一个新的 RS，然后逐渐将新 RS 中的副本数增加到理想状态，将旧 RS 中的副本数减小到 0 的复合操作。

（6）无论是否理想，Kubernetes 最终都会使应用程序全部更新到期望状态，并保持最大的副本总数和可用副本总数的不变性。

第 6 章 Kubernetes+NFS 持久化存储实战

6.1 Kubernetes 服务运行状态

Kubernetes 运行的服务，从简单到复杂可以分成三类：无状态服务、普通有状态服务和有状态集群服务。下面分别介绍 Kubernetes 是如何运行这三类服务的。

（1）无状态服务。

Kubernetes 使用 RC（或更新的 ReplicaSet）保证一个服务的实例数量，如果某个 Pod 实例由于某种原因崩溃了，RC 会立刻用这个 Pod 的模板启动一个 Pod 替代它，由于是无状态服务，新启动的 Pod 与原来健康状态下的 Pod 一模一样。在 Pod 被重建后，它的 IP 地址可能发生变化，为了对外提供一个稳定的访问接口，Kubernetes 引入了 service。一个 service 后面可以挂多个 Pod，从而实现服务的高可用。

（2）普通有状态服务。

与无状态服务相比，它多了状态保存的需求。Kubernetes 提供了以 Volume 和 Persistent Volume 为基础的存储系统，可以实现服务的状态保存。

（3）有状态集群服务。

与普通有状态服务相比，它多了集群管理的需求。Kubernetes 为此开发了一套以 PetSet 为核心的全新特性，方便了有状态集群服务在 Kubernetes 上的部署和管理。具体来说是通过 Init Container 做集群的初始化工作，用无头服务（Headless Service）维持集群成员的稳定关系，用动态存储供给方便集群扩容，最后用 PetSet 综合管理整个集群。

6.2 Kubernetes 存储系统

Kubernetes 的存储系统从基础到高级又大致分为三个层次：普通存储卷、持久存储卷（Persistent Volume）和动态存储供应（Dynamic Provisioning）。

1）普通存储卷之单节点存储卷

单节点存储卷是最简单的普通存储卷，它和 Docker 的存储卷类似，使用的是 Pod 所在 Kubernetes 节点的本地目录。具体有两种，一种是 emptyDir，是一个匿名的空目录，由 Kubernetes 在创建 Pod 时创建，删除 Pod 时删除；另外一种是 hostPath，与 emptyDir 的区别在于，后者在 Pod 之外独立存在，由用户指定路径名。这类和节点绑定的存储卷在 Pod 迁移到其他节点后数据就会丢失，所以只能用于存储临时数据或用于在同一个 Pod 里的容器之间共享数据。普通 Volume 目前支持的各种存储插件及情况如图 6-1 所示。

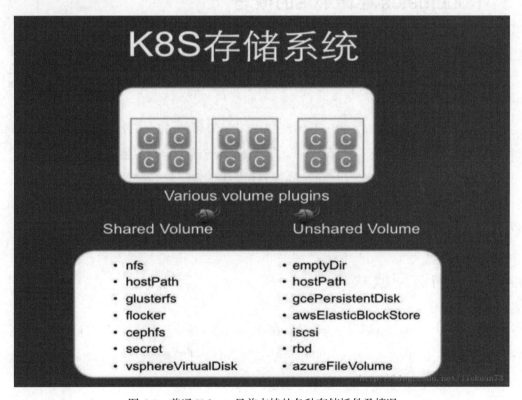

图 6-1 普通 Volume 目前支持的各种存储插件及情况

2）持久存储卷

普通存储卷和使用它的 Pod 之间是一种静态绑定关系，在定义 Pod 的文件里，同时定义了 Pod 使用的存储卷。存储卷是 Pod 的附属品，无法单独创建一个存储卷，因为它不是一个独立的 Kubernetes 资源对象。

而持久存储卷（Persistent Volume，PV）是一个 Kubernetes 资源对象，所以可以单独创建一个 PV。它不和 Pod 直接发生关系，而是通过 Persistent Volume Claim（PV 索取，PVC）实现动态绑定。Pod 定义里指定的是 PVC，然后 PVC 会根据 Pod 的要求自动绑定合适的 PV 给 Pod 使用。

3）动态存储供应

与普通存储卷类似，这里不再赘述。

6.3　Kubernetes 存储绑定的概念

用户根据所需存储空间大小和访问模式创建（或在动态部署中已创建）一个 PVC。

Kubernetes 的 master 节点循环监控新产生的 PVC，找到与之匹配的 PV（如果有），并把它们绑定在一起。

动态配置时，循环会一直将 PV 与这个 PVC 绑定，直到 PV 完全匹配 PVC，以避免 PVC 请求和得到的 PV 不一致。绑定一旦形成，PVC 绑定就是专有的，不管是使用何种模式绑定的。

如果找不到匹配的存储卷，用户请求会一直保持未绑定状态。在匹配的存储卷可用之后，用户请求将会被绑定。例如，一个配置很多 50Gi PV 的集群不会匹配到一个要求 100Gi 的 PVC。只有在 100Gi PV 被加到集群之后，这个 PVC 才可以被绑定。

6.4　PV 的访问模式

Kubernetes PV 访问模式通常分为 3 种：ReadWriteOnce 是最基本的方式，可读可写，但只支持被单个 Pod 挂载；ReadOnlyMany 以只读的方式被多个 Pod 挂载；ReadWriteMany 可以以读写的方式被多个 Pod 共享。不是每一种存储都支持这 3 种方式。

在 PVC 绑定 PV 时通常根据两个条件绑定：一个是存储的大小，另一个就是访问模式。Kubernetes PV 种类和访问类型如图 6-2 所示。访问模式全称及简写如下。

（1）ReadWriteOnce：RWO。

（2）ReadOnlyMany：ROX。

（3）ReadWriteMany：RWX。

Volume Plugin	ReadWriteOnce	ReadOnlyMany	ReadWriteMany
AWSElasticBlockStore	✓	-	-
AzureFile	✓	✓	✓
AzureDisk	✓	-	-
CephFS	✓	✓	✓
Cinder	✓	-	-
FC	✓	✓	-
FlexVolume	✓	✓	-
Flocker	✓	-	-
GCEPersistentDisk	✓	✓	-
Glusterfs	✓	✓	✓
HostPath	✓	-	-
iSCSI	✓	✓	-
PhotonPersistentDisk	✓	-	-
Quobyte	✓	✓	✓
NFS	✓	✓	✓
RBD	✓	✓	-

图 6-2　Kubernetes PV 种类和访问类型

PV 的生命周期首先是 Provision，即创建 PV。这里创建 PV 有两种方式，分别为静态创建和动态创建。

（1）静态创建：管理员手动创建一批 PV，组成一个 PV 池，供 PVC 绑定。

（2）动态创建：在现有 PV 不满足 PVC 的请求时，可以使用存储分类（StorageClass）。PV 先创建分类，PVC 请求已创建的某个类（StorageClass）的资源，这样就达到动态配置的效果，即通过一个叫 StorageClass 的对象由存储系统根据 PVC 的要求自动创建。

PV 创建完后状态会变成 Available，等待被 PVC 绑定。一旦被 PVC 邦定，PV 的状态会变成 Bound，就可以被定义了相应 PVC 的 Pod 使用。Pod 使用完后会释放 PV，PV 的状态变成 Released。变成 Released 的 PV 会根据定义的回收策略做相应的回收工作。

有 3 种回收策略：Retain、Delete 和 Recycle。Retain 策略就是保留现场，Kubernetes 什么

也不做，等待用户手动处理 PV 中的数据，处理完后，再手动删除 PV。Delete 策略，Kubernetes 会自动删除该 PV 及其中的数据。Recycle 策略，Kubernetes 会将 PV 中的数据删除，然后把 PV 的状态变成 Available，又可以被新的 PVC 绑定使用。

在实际的使用场景中，PV 的创建和使用通常不是同一个用户完成。这里有一个典型的应用场景：管理员创建一个 PV 池，开发人员创建 Pod 和 PVC，PVC 里定义了 Pod 所需存储的大小和访问模式，然后 PVC 会到 PV 池里自动匹配最合适的 PV 给 Pod 使用。

6.5 Kubernetes+NFS 静态存储模式

下面为静态 PV 存储操作方式。

基于 Linux 平台构建 NFS 网络文件系统，操作指令如下：

```
#安装 NFS 文件服务
yum install nfs-utils -y
#配置共享目录和权限
vim /etc/exports
/data/ *(rw,async,no_root_squash)
#启动 NFS 服务
service nfs restart
```

创建 PV 文件，pv.yaml 文件内容如下：

```
cat>pv.yaml<<EOF
apiVersion: v1
kind: PersistentVolume
metadata:
  name: nfs-pv
  namespace: default
spec:
  capacity:
    storage: 10G
  accessModes:
    - ReadWriteMany
  nfs:
    # FIXME: use the right IP
    server: 192.168.1.147
    path: /data/
EOF
```

PV 配置参数如下：

```
Capacity 指定 PV 的容量为 100MB。
accessModes 指定访问模式为 ReadWriteOnce,支持的访问模式有：
ReadWriteOnce - PV 能以 read-write 模式挂载到单个节点。
ReadOnlyMany - PV 能以 read-only 模式挂载到多个节点。
ReadWriteMany - PV 能以 read-write 模式挂载到多个节点。
persistentVolumeReclaimPolicy 指定当 PV 的回收策略为 Recycle,支持的策略有：
Retain - 需要管理员手动回收。
Recycle - 清除 PV 中的数据,效果相当于执行 rm -rf /thevolume/*。
Delete - 删除 Storage Provider 上的对应存储资源,如 AWS EBS、GCE PD、Azure、Disk、
OpenStack Cinder Volume 等。
storageClassName 指定 PV 的 class 为 NFS。相当于为 PV 设置了一个分类,PVC 可以指
定 class 申请相应 class 的 PV。
指定 PV 在 NFS 服务器上对应的目录。
```

6.6 PVC 存储卷创建

（1）创建 PVC，pvc.yaml 文件内容如下：

```
cat>pvc.yaml<<EOF
apiVersion: v1
kind: PersistentVolumeClaim
metadata:
  name: nfs-pvc
  namespace: default
spec:
  accessModes:
    - ReadWriteMany
  storageClassName: ""
  resources:
    requests:
      storage: 10G
EOF
```

（2）创建 PV、PVC 之后，需要通过 kubectl apply –f 读取 YAML 配置，使其部署生效。最终效果如图 6-3 所示。

图 6-3　Kubernetes PV、PVC 案例实战

6.7　Nginx 整合 PV 存储卷

（1）创建 Nginx Pod 容器，使用 PVC，nginx.yaml 文件内容如下：

```
cat>nginx.yaml<<EOF
apiVersion: v1
kind: ReplicationController
metadata:
  name: nginx-v1
  labels:
    name: nginx-v1
  namespace: default
spec:
  replicas: 1
  selector:
    name: nginx-v1
  template:
    metadata:
      labels:
        name: nginx-v1
    spec:
      containers:
      - name: nginx-v1
        image: nginx
        volumeMounts:
        - mountPath: /usr/share/nginx/html
          name: nginx-data
```

```
        ports:
        - containerPort: 80
      volumes:
      - name: nginx-data
        persistentVolumeClaim:
          claimName: nfs-pvc
EOF
```

（2）登录 node 节点，查看 NFS PV 资源是否挂载，如图 6-4 所示。

```
[root@localhost ~]# mount|grep nfs
192.168.0.151:/data on /var/lib/kubelet/pods/4aac8416-5a
s.io~nfs/pv type nfs4 (rw,relatime,vers=4.1,rsize=131072
t=0,timeo=600,retrans=2,sec=sys,clientaddr=192.168.0.142
[root@localhost ~]#
[root@localhost ~]# docker ps
CONTAINER ID              IMAGE
        CREATED                 STATUS                    PORTS
34e815c2b48a              nginx
..."        3 minutes ago           Up 3 minutes
sj4c_default_4aac8416-5ac0-11e9-9d56-000c292c5e36_50bc6
8c25dafb09e8              registry.access.redhat.com/rhel7/po
```

图 6-4　Kubernetes PV、PVC 案例测试 1

（3）测试 Kubernetes Pod NFS 与 NFS 服务器数据是否一致，如图 6-5 所示。

```
命令行 nginx-v1         ▼  在 nginx-v1-cc7zf
root@nginx-v1-cc7zf:~# cd
root@nginx-v1-cc7zf:~#
root@nginx-v1-cc7zf:~# cd /usr/share/nginx/html/
root@nginx-v1-cc7zf:/usr/share/nginx/html#
root@nginx-v1-cc7zf:/usr/share/nginx/html# ls
backup  etcd  index.html  www.jfedu.net
root@nginx-v1-cc7zf:/usr/share/nginx/html#
root@nginx-v1-cc7zf:/usr/share/nginx/html# echo Test POD NFS Pages > index.html
root@nginx-v1-cc7zf:/usr/share/nginx/html#
root@nginx-v1-cc7zf:/usr/share/nginx/html#
root@nginx-v1-cc7zf:/usr/share/nginx/html# cat index.html
Test POD NFS Pages
root@nginx-v1-cc7zf:/usr/share/nginx/html#
```

(a)

图 6-5　Kubernetes PV、PVC 案例测试 2

```
root         4131     2  0 10:41 ?        00:00:00 [nfsd]
root         4132     2  0 10:41 ?        00:00:00 [nfsd]
root         4133     2  0 10:41 ?        00:00:00 [nfsd]
root         4134     2  0 10:41 ?        00:00:00 [nfsd]
root         4156  1361  0 10:52 pts/0    00:00:00 grep -
[root@jfedu141 ~]# cd /data/
[root@jfedu141 data]# ls
etcd   index.html
[root@jfedu141 data]#
[root@jfedu141 data]# cat index.html
Test POD NFS Pages
[root@jfedu141 data]#
```

（b）

图 6-5 （续）

6.8 Kubernetes+NFS 动态存储模式

动态创建 PV，是指在现有 PV 不满足 PVC 的请求时，可以使用存储分类（StorageClass），描述具体过程：PV 先创建分类，PVC 请求已创建的某个类（StorageClass）的资源，这样就达到动态配置的效果。即通过一个叫 StorageClass 的对象由存储系统根据 PVC 的要求自动创建。

其中动态方式是通过 StorageClass 完成的，这是一种新的存储供应方式。动态卷供给能力让管理员不必进行预先创建存储卷，而是随用户需求而创建，如图 6-6 所示。

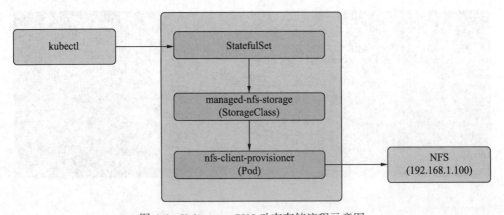

图 6-6 Kubernetes PVC 动态存储流程示意图

使用 StorageClass 有什么好处呢？StorageClass 除了由存储系统动态创建，节省了管理员的时间外，还可以封装不同类型的存储供 PVC 选用。

在 StorageClass 出现以前，PVC 绑定一个 PV 只能根据两个条件，一个是存储的大小，另一个是访问模式。在 StorageClass 出现后，等于增加了一个绑定维度。以下为动态 PV 操作方式。

基于 Linux 平台构建 NFS 网络文件系统，配置指令如下：

```
#安装 NFS 文件服务
yum install nfs-utils -y
#配置共享目录和权限
vim /etc/exports
/data/ *(rw,async,no_root_squash)
#启动 NFS 服务
service nfs restart
```

6.9 NFS 插件配置实战

（1）NFS 默认不支持动态存储，而是使用了第三方的 NFS 插件安装 NFS 插件，GitHub 地址如下：

```
https://github.com/kubernetes-incubator/external-storage/tree/master/nfs-client/deploy
```

（2）下载 NFS 和动态 PV 配置文件，操作指令如下：

```
for file in class.yaml deployment.yaml rbac.yaml ; do wget -c https://raw.githubusercontent.com/kubernetes-incubator/external-storage/master/nfs-client/deploy/$file ; done
```

（3）修改 deployment.yaml 文件内容，将 NFS 地址和路径修改正确，如图 6-7 所示。

```
    containers:
      - name: nfs-client-provisioner
        image: quay.io/external_storage/nfs-client-provisioner:latest
        volumeMounts:
          - name: nfs-client-root
            mountPath: /persistentvolumes
        env:
          - name: PROVISIONER_NAME
            value: fuseim.pri/ifs
          - name: NFS_SERVER
            value: 192.168.1.147
          - name: NFS_PATH
            value: /data
    volumes:
      - name: nfs-client-root
        nfs:
          server: 192.168.1.147
          path: /data
```

图 6-7　Kubernetes PVC 动态存储 YAML 文件修改

（4）依次应用 rbac.yaml、class.yaml 和 deployment.yaml 配置文件，操作指令如下：

```
kubectl create -f rbac.yaml
kubectl create -f class.yaml
kubectl create -f deployment.yaml
```

（5）最后创建 Kubernetes Pod 案例，自动获取 PV 资源即可，Pod 案例 YAML 文件代码如下：

```yaml
apiVersion: v1
kind: Service
metadata:
  name: nginx
  labels:
    app: nginx
spec:
  ports:
  - port: 80
    name: nginx
  selector:
    app: nginx
  clusterIP: None
apiVersion: apps/v1
kind: StatefulSet
metadata:
  name: web
spec:
  selector:
    matchLabels:
      app: nginx
  serviceName: "nginx"
  replicas: 1
  template:
    metadata:
      labels:
        app: nginx
    spec:
      imagePullSecrets:
      - name: huoban-harbor
      terminationGracePeriodSeconds: 10
      containers:
      - name: nginx
        image: nginx
```

```
      ports:
      - containerPort: 80
        name: web
      volumeMounts:
      - name: www
        mountPath: /usr/share/nginx/html
  volumeClaimTemplates:
  - metadata:
      name: www
    spec:
      accessModes: [ "ReadWriteOnce" ]
      storageClassName: "managed-nfs-storage"
      resources:
        requests:
          storage: 1Gi
```

（6）根据以上方法，如果 Pod 容器一直处于 Pending（挂起）状态，可以查看其 LOG 日志信息，报错信息为 Claim reference: selfLink was empty, can't make refere，解决方法如下：

```
#在/etc/kubernetes/manifests/kube-apiserver.yaml 配置文件中加入如下代码
- --feature-gates=RemoveSelfLink=false
#重新应用 apiserver.yaml 文件,同时重新创建所有 NFS yaml 文件即可
kubectl apply -f /etc/kubernetes/manifests/kube-apiserver.yaml
```

（7）在 NFS /data/目录下，能够看到 default 开头的动态 PV 创建的目录和相关的文件即可。

（8）查看 Nginx Pod 对应的 service，显示没有 Cluster IP，如图 6-8 所示。

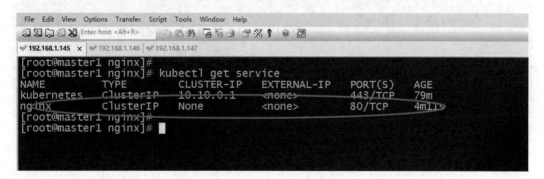

图 6-8　Kubernetes Pod service 案例实战 1

（9）验证解析，每个 Pod 都拥有一个基于其顺序索引的、稳定的主机名，如图 6-9 所示，操作指令如下：

```
kubectl get pods --namespace=default
```

```
kubectl exec -it web-0 --namespace default /bin/bash
hostname
```

```
[root@master1 nginx]#
[root@master1 nginx]# cd
[root@master1 ~]#
[root@master1 ~]#
[root@master1 ~]# kubectl get pods --namespace=default
NAME                                        READY   STATUS    RESTARTS   AGE
nfs-client-provisioner-b5cb7bdf8-g2vd5      1/1     Running   1          46m
web-0                                       1/1     Running   0          10m
[root@master1 ~]#
```

图 6-9　Kubernetes Pod service 案例实战 2

（10）基于 kubectl run 运行一个提供 nslookup 命令的容器，该命令来自于 dnsutils 包。通过对 Pod 的主机名执行 nslookup 命令，可以检查它们在集群内部的 DNS 地址，操作指令如下，结果如图 6-10 所示。

```
kubectl run -i --tty --image busybox:1.28.4 -n default dns-test --restart=Never --rm
nslookup web-0.nginx
```

```
[root@master1 ~]# kubectl run -i --tty --image busybox:1.28.4 -n default dns-test --r
If you don't see a command prompt, try pressing enter.
/ #
/ # nslookup web-0.nginx
Server:    10.10.0.10
Address 1: 10.10.0.10 kube-dns.kube-system.svc.cluster.local

Name:      web-0.nginx
Address 1: 10.244.1.9 web-0.nginx.default.svc.cluster.local
/ #
/ # ping -c 3 web-0.nginx.default.svc.cluster.local
PING web-0.nginx.default.svc.cluster.local (10.244.1.9): 56 data bytes
64 bytes from 10.244.1.9: seq=0 ttl=64 time=0.131 ms
64 bytes from 10.244.1.9: seq=1 ttl=64 time=0.109 ms
64 bytes from 10.244.1.9: seq=2 ttl=64 time=0.109 ms
```

图 6-10　Kubernetes Pod 案例测试 1

（11）进入容器内部，访问 web-0.nginx.default.svc.cluster.local 域名，操作指令如下，结果如图 6-11 所示。

```
curl web-0.nginx.default.svc.cluster.local
```

```
[root@node1 ~]# docker ps|grep nginx
d5fa2a54dbeb        nginx                                              "/docker-entr
lt_f319c35e-e558-45e4-bcf3-59bd47c97094_0
[root@node1 ~]#
[root@node1 ~]# docker exec -it d5fa2a54dbeb /bin/bash
root@web-0:/#
root@web-0:/#
root@web-0:/# cat /etc/resolv.conf
nameserver 10.10.0.10
search default.svc.cluster.local svc.cluster.local cluster.local localdomain
options ndots:5
root@web-0:/#
root@web-0:/# ping -c3 web-0.nginx.default.svc.cluster.local
bash: ping: command not found
root@web-0:/#
root@web-0:/# curl web-0.nginx.default.svc.cluster.local
www.jfedu.net test page
root@web-0:/#
```

图 6-11　Kubernetes Pod 案例测试 2

（12）删除 web-0 Pod 容器，然后系统会自动创建一台，此时再次查看其域名对应的 IP 地址，如果通过 IP 访问还是访问到之前内容，即证明持久化操作成功，如图 6-12 所示。

```
[root@master1 ~]#
[root@master1 ~]# kubectl run -i --tty --image busybox:1.28.4 -n default dns-test --restart=Never --rm
If you don't see a command prompt, try pressing enter.
/ #
/ #
/ # nslookup web-0.nginx
Server:    10.10.0.10
Address 1: 10.10.0.10 kube-dns.kube-system.svc.cluster.local

Name:      web-0.nginx
Address 1: 10.244.1.15 web-0.nginx.default.svc.cluster.local
/ #
/ # exit
pod "dns-test" deleted
[root@master1 ~]# curl 10.244.1.15
www.jfedu.net test page
[root@master1 ~]#
```

图 6-12　Kubernetes Pod 案例测试 3

不管 web-0 重新调度去哪个 node 上，它都会继续监听各自的主机名，因为与其 PVC 相关联的 PV 被重新挂载到了对应的 VolumeMount 上。其 PV 将会被挂载到合适的挂载点上。

第 7 章 Kubernetes+CephFS 持久化存储实战

CephFS 模式下的 Kubernetes 服务运行状态、Kubernetes 存储系统、Kubernetes 存储绑定的概念、PV 的访问模式同 NFS 模式下，见 6.1 节～6.4 节。

7.1 Kubernetes+CephFS 静态存储模式

Kubernetes 使用 CephFS 共享静态存储模式，需要先创建静态 PV，再手动创建 PVC，同时 PVC 绑定 PV 之后，方可创建部署业务使用 PV 资源。

7.2 PV 存储卷创建

（1）创建 Kubernetes CephFS 密钥，操作指令如下：

```
ceph auth get-key client.admin > /tmp/secret
kubectl create namespace cephfs
kubectl create secret generic ceph-admin-secret --from-file=/tmp/secret
```

（2）创建 PV，创建 pv.yaml 文件，操作指令如下：

```
cat>pv.yaml<<EOF
apiVersion: v1
kind: PersistentVolume
metadata:
  name: cephfs-pv1
spec:
  capacity:
    storage: 1Gi
```

```
  accessModes:
    - ReadWriteMany
  cephfs:
    monitors:
      - 192.168.1.145:6789
    user: admin
    secretRef:
      name: ceph-admin-secret
    readOnly: false
  persistentVolumeReclaimPolicy: Recycle
EOF
```

(3) Kubernetes PV 配置参数如下:

```
Capacity 指定 PV 的容量为 100MB。
accessModes 指定访问模式为 ReadWriteOnce,支持的访问模式有:
ReadWriteOnce - PV 能以 read-write 模式挂载到单个节点。
ReadOnlyMany - PV 能以 read-only 模式挂载到多个节点。
ReadWriteMany - PV 能以 read-write 模式挂载到多个节点。
persistentVolumeReclaimPolicy 指定当 PV 的回收策略为 Recycle,支持的策略有:
Retain - 需要管理员手动回收。
Recycle - 清除 PV 中的数据,效果相当于执行 rm -rf /thevolume/*。
Delete - 删除 Storage Provider 上的对应存储资源,如 AWS EBS、GCE PD、Azure、Disk、
OpenStack Cinder Volume 等。
storageClassName 指定 PV 的 class 为 NFS。相当于为 PV 设置了一个分类,PVC 可以指
定 class 申请相应 class 的 PV。
指定 PV 在 NFS 服务器上对应的目录。
```

7.3 PVC 存储卷创建

Kubernetes CephFS PV 资源创建如图 7-1 所示。pvc.yaml 文件内容如下:

```
cat>pvc.yaml<<EOF
kind: PersistentVolumeClaim
apiVersion: v1
metadata:
  name: cephfs-pv-claim1
spec:
  accessModes:
    - ReadWriteMany
```

```
      resources:
        requests:
          storage: 1Gi
    EOF
```

集群	元数据			
Cluster Roles				
Namespaces	名字	创建时间	经过的时间	UID
Nodes	cephfs-pv1	2021年7月27日	21 minutes	8e610b20-41f9-4edb-95d4-7a51c07eff65
Persistent Volumes	注释			
Storage Classes	pv.kubernetes.io/bound-by-controller: yes			
命名空间	资源信息			
default	状态	要求	回收策略	
概况	Bound	default/cephfs-pv-claim1	Recycle	

图 7-1 Kubernetes CephFS PV 资源创建

7.4 Nginx 整合 CephFS PV 存储卷

（1）创建 Nginx Pod 容器使用 PVC，nginx.yaml 文件内容如下：

```
cat>nginx.yaml<<EOF
apiVersion: v1
kind: ReplicationController
metadata:
  name: nginx-v1
  labels:
    name: nginx-v1
  namespace: default
spec:
  replicas: 1
  selector:
    name: nginx-v1
  template:
    metadata:
      labels:
        name: nginx-v1
    spec:
      containers:
```

```
      - name: nginx-v1
        image: nginx
        volumeMounts:
        - mountPath: /usr/share/nginx/html
          name: nginx-data
        ports:
        - containerPort: 80
      volumes:
      - name: nginx-data
        persistentVolumeClaim:
          claimName: cephfs-pv-claim1
EOF
```

（2）登录 node 节点，查看 CephFS PV 资源是否挂载，如图 7-2 所示。

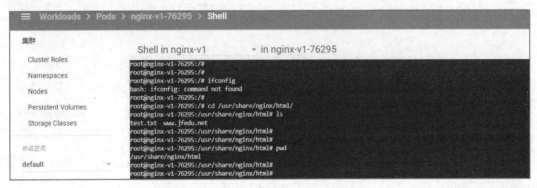

(a)

(b)

图 7-2　Kubernetes CephFS PV 存储实战 1

（3）测试 Kubernetes Pod CephFS 和 CephFS 服务器数据是否一致，如图 7-3 所示。

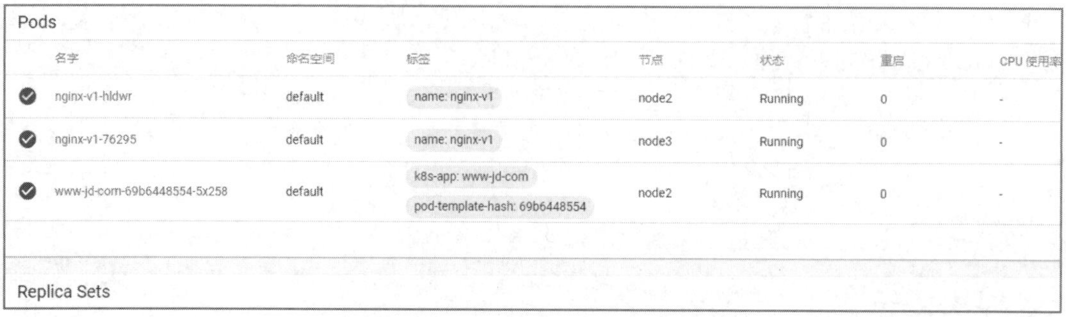

Pods						
名字	命名空间	标签	节点	状态	重启	CPU 使用率
✓ nginx-v1-hldwr	default	name: nginx-v1	node2	Running	0	-
✓ nginx-v1-76295	default	name: nginx-v1	node3	Running	0	-
✓ www-jd-com-69b6448554-5x258	default	k8s-app: www-jd-com pod-template-hash: 69b6448554	node2	Running	0	-

Replica Sets

（a）

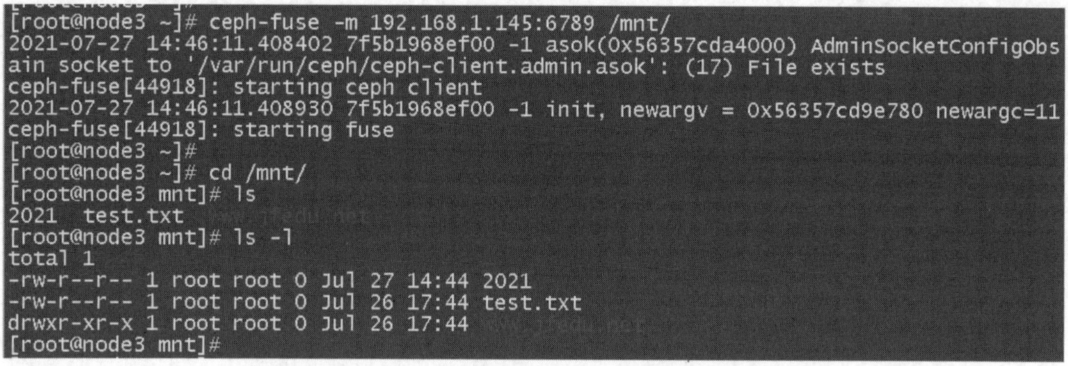

```
[root@node3 ~]# ceph-fuse -m 192.168.1.145:6789 /mnt/
2021-07-27 14:46:11.408402 7f5b1968ef00 -1 asok(0x56357cda4000) AdminSocketConfigObs
ain socket to '/var/run/ceph/ceph-client.admin.asok': (17) File exists
ceph-fuse[44918]: starting ceph client
2021-07-27 14:46:11.408930 7f5b1968ef00 -1 init, newargv = 0x56357cd9e780 newargc=11
ceph-fuse[44918]: starting fuse
[root@node3 ~]#
[root@node3 ~]# cd /mnt/
[root@node3 mnt]# ls
2021  test.txt
[root@node3 mnt]# ls -l
total 1
-rw-r--r-- 1 root root 0 Jul 27 14:44 2021
-rw-r--r-- 1 root root 0 Jul 26 17:44 test.txt
drwxr-xr-x 1 root root 0 Jul 26 17:44
[root@node3 mnt]#
```

（b）

图 7-3　Kubernetes CephFS PV 存储实战 2

7.5　Kubernetes+CephFS 动态存储模式

　　Kubernetes 使用 CephFS 共享动态存储模式，需要动态创建 PV，是指在现有 PV 不满足 PVC 的请求时，可以使用存储分类（StorageClass），描述具体过程为：PV 先创建分类，PVC 请求已创建的某个类（StorageClass）的资源，这样就达到动态配置的效果。即通过一个叫 StorageClass 的对象由存储系统根据 PVC 的要求自动创建。

　　其中动态方式是通过 StorageClass 完成的，这是一种新的存储供应方式。动态卷供给能力让管理员不必进行预先创建存储卷，而是随用户需求进行创建，如图 7-4 所示。

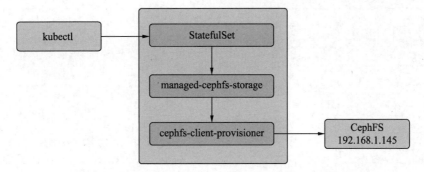

图 7-4 Kubernetes CephFS PVC 动态存储流程示意图

使用 StorageClass 有什么好处呢？除了由存储系统动态创建，节省了管理员的时间外，还可以封装不同类型的存储供 PVC 选用。

在 StorageClass 出现以前，PVC 绑定一个 PV 只能根据两个条件，一个是存储的大小，另一个是访问模式。在 StorageClass 出现后，等于增加了一个绑定维度。

7.6 CephFS 动态插件配置实战

（1）CephFS 默认不支持动态存储，而是使用第三方的 Cephfs-provisioner 插件安装 Cephfs-provisioner 插件，GitHub 地址如下：

```
https://kubernetes.io/docs/concepts/storage/storage-classes/
```

（2）下载 Cephfs-provisioner 与动态 PV 配置文件。

```
#CephFS 需要使用两个 Pool 来分别存储数据和元数据
ceph osd pool create fs_data 128
ceph osd pool create fs_metadata 128
ceph osd lspools
#创建一个 CephFS
ceph fs new cephfs fs_metadata fs_data
#查看 CephFS
ceph fs ls
```

（3）执行以上 YAML 文件，使其生效，并查看其 Pod 信息。

```
kubectl apply -f external-storage-cephfs-provisioner.yaml
```

（4）查看 Pod 状态，如图 7-5 所示，操作命令如下：

```
kubectl get pod -n kube-system |grep -aiE provisioner
```

Cluster Roles			
Namespaces	✓	cephfs-provisioner	
Nodes	✓	coredns	k8s-app: kube-dns
Persistent Volumes			
Storage Classes			

命名空间
kube-system

Pods

	名字	标签	节点
✓	cephfs-provisioner-6cd5548-2cdjh	app: cephfs-provisioner pod-template-hash: 6cd5548	node2

图 7-5 Kubernetes CephFS PVC 动态存储实战

（5）创建 CephFS secret 密钥，操作指令如下：

```
#查看 key 在 ceph 的 mon 或者 admin 节点
ceph auth get-key client.admin
ceph auth get-key client.kube
#创建 admin secret
#将 CEPH_ADMIN_SECRET 替换为 client.admin 获取到的 key
export CEPH_ADMIN_SECRET='AQCXgP5giXmCARAAzz1mWKqJ+dTFdLArr1Ee+Q=='
kubectl create secret generic ceph-secret --type="kubernetes.io/rbd" \
--from-literal=key=$CEPH_ADMIN_SECRET \
--namespace=kube-system
#查看 secret
kubectl get secret ceph-user-secret -o yaml
kubectl get secret ceph-secret -n kube-system -o yaml
```

（6）配置 StorageClass，操作方法如下：

```
#如果使用 kubeadm 创建的集群 provisioner,则使用如下方式
# provisioner: ceph.com/rbd
cat >storageclass-cephfs.yaml<<EOF
kind: StorageClass
apiVersion: storage.Kubernetes.io/v1
metadata:
  name: dynamic-cephfs
provisioner: ceph.com/cephfs
parameters:
    monitors: 192.168.1.145:6789
    adminId: admin
    adminSecretName: ceph-secret
    adminSecretNamespace: "kube-system"
    claimRoot: /volumes/kubernetes
```

```
EOF
kubectl apply -f storageclass-ceph-rdb.yaml
kubectl get sc
```

（7）创建 CephFS PVC 资源索取，操作指令如下：

```
cat>cephfs-pvc.yaml<<EOF
kind: PersistentVolumeClaim
apiVersion: v1
metadata:
  name: cephfs-claim
spec:
  accessModes:
    - ReadWriteOnce
  storageClassName: dynamic-cephfs
  resources:
    requests:
      storage: 2Gi
EOF
kubectl apply -f cephfs-pvc.yaml
```

（8）创建 Nginx Pod 容器，使用 CephFS 资源池，如图 7-6 所示，操作指令如下：

```
cat>nginx-pod.yaml<<EOF
apiVersion: v1
kind: Pod
metadata:
  name: nginx-pod1
  labels:
    name: nginx-pod1
spec:
  containers:
  - name: nginx-pod1
    image: docker.io/library/nginx:latest
    ports:
    - name: web
      containerPort: 80
    volumeMounts:
    - name: cephfs
      mountPath: /usr/share/nginx/html
  volumes:
  - name: cephfs
    persistentVolumeClaim:
      claimName: cephfs-claim
EOF
```

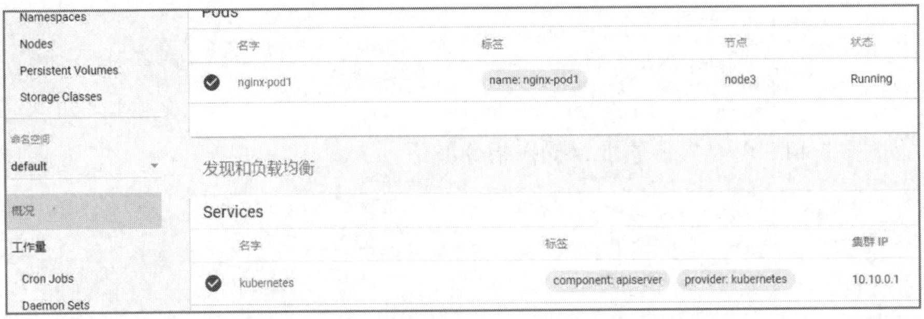

（a）

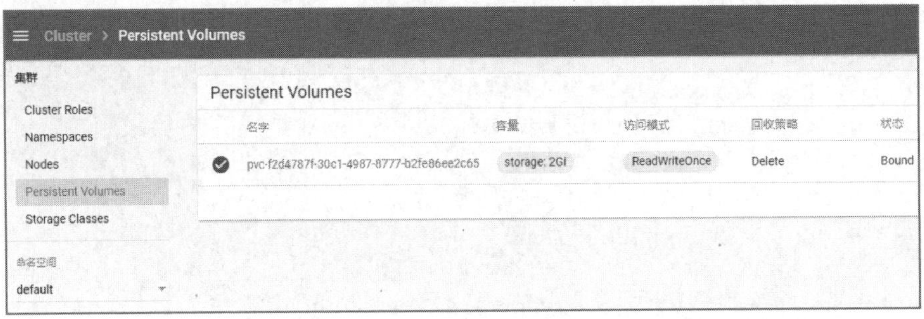

（b）

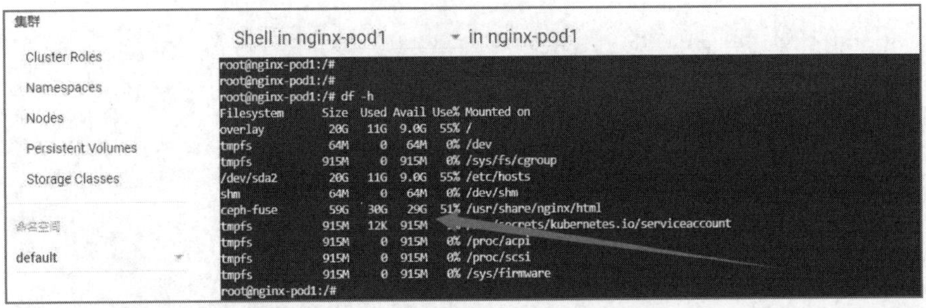

（c）

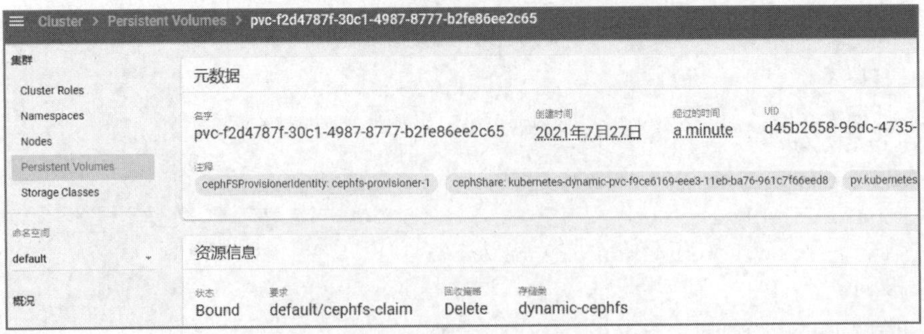

（d）

图 7-6　Kubernetes CephFS PVC 动态存储实战

第 8 章 Kubernetes+Ceph RBD 持久化存储实战

Ceph RBD 模式下的 Kubernetes 服务运行状态、Kubernetes 存储系统、Kubernetes 存储绑定概念、PV 访问模式同 NFS 模式和 CephFS 模式下，详见 6.1 节～6.4 节。

8.1 Kubernetes+Ceph RBD 静态存储模式

Kubernetes 使用 Ceph RBD 共享静态存储模式，需要先创建静态 PV，再手动创建 PVC，同时 PVC 绑定 PV 之后，方可创建部署业务使用 PV 资源。

8.2 PV 存储卷创建

（1）创建 Kubernetes Ceph RBD 密钥，操作指令如下：

```
ceph auth get-key client.admin > /tmp/secret
kubectl create namespace ceph rbd
kubectl create secret generic ceph-admin-secret --from-file=/tmp/secret
```

（2）创建 Ceph pool 和 Image，操作指令如下：

```
ceph osd pool create kube-nginx 128 128
rbd create kube-nginx/rbd0 -s 10G --image-feature layering
```

（3）创建 PV，pv.yaml 文件内容如下：

```
cat>pv.yaml<<EOF
apiVersion: v1
kind: PersistentVolume
```

```
metadata:
  name: rbd-pv1
spec:
  capacity:
    storage: 1Gi
  accessModes:
    - ReadWriteOnce
  rbd:
    monitors:
      - 192.168.1.145:6789
    pool: kube-nginx
    image: rbd0
    user: admin
    secretRef:
      name: ceph-admin-secret
  persistentVolumeReclaimPolicy: Recycle
EOF
```

（4）PV 配置参数如下：

```
Capacity 指定 PV 的容量为 100MB。
accessModes 指定访问模式为 ReadWriteOnce,支持的访问模式有：
ReadWriteOnce - PV 能以 read-write 模式挂载到单个节点。
ReadOnlyMany - PV 能以 read-only 模式挂载到多个节点。
ReadWriteMany - PV 能以 read-write 模式挂载到多个节点。
persistentVolumeReclaimPolicy 指定当 PV 的回收策略为 Recycle,支持的策略有：
Retain - 需要管理员手动回收。
Recycle - 清除 PV 中的数据,效果相当于执行 rm -rf /thevolume/*。
Delete - 删除 Storage Provider 上的对应存储资源,如 AWS EBS、GCE PD、Azure、Disk、
OpenStack Cinder Volume 等。
storageClassName 指定 PV 的 class 为 NFS。相当于为 PV 设置了一个分类,PVC 可以指
定 class 申请相应 class 的 PV。
指定 PV 在 NFS 服务器上对应的目录。
```

8.3 PVC 存储卷创建

Kubernetes Ceph RBD PV 创建如图 8-1 所示，pvc.yaml 文件内容如下：

```
cat>pvc.yaml<<EOF
kind: PersistentVolumeClaim
apiVersion: v1
metadata:
  name: rbd-pv-claim1
spec:
  accessModes:
    - ReadWriteOnce
  resources:
    requests:
      storage: 1Gi
EOF
```

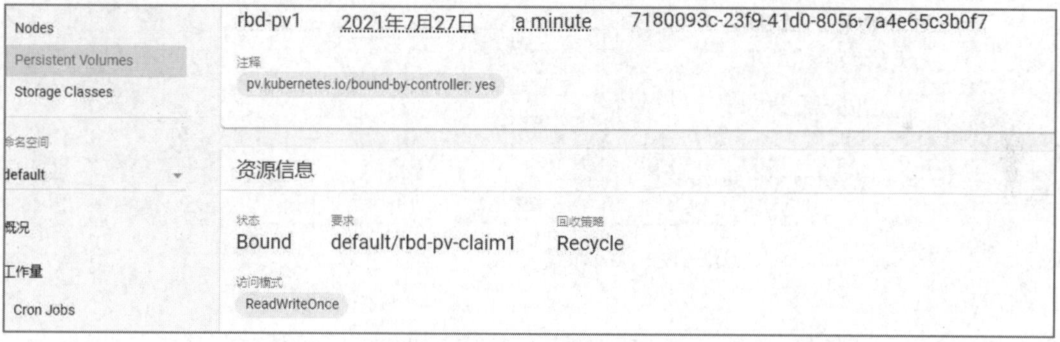

图 8-1 Kubernetes Ceph RBD PV 创建

8.4 Nginx 整合 Ceph PV 存储卷

（1）创建 Nginx Pod 容器使用 PVC，nginx.yaml 文件内容如下：

```
cat>nginx.yaml<<EOF
apiVersion: v1
kind: ReplicationController
metadata:
  name: nginx-v2
```

```
      labels:
        name: nginx-v2
      namespace: default
    spec:
      replicas: 1
      selector:
        name: nginx-v2
      template:
        metadata:
          labels:
            name: nginx-v2
        spec:
          containers:
          - name: nginx-v2
            image: nginx
            volumeMounts:
            - mountPath: /usr/share/nginx/html
              name: nginx-data
            ports:
            - containerPort: 80
          volumes:
          - name: nginx-data
            persistentVolumeClaim:
              claimName: rbd-pv-claim1
EOF
```

（2）登录 node 节点，查看 Ceph PV 资源是否挂载，如图 8-2 所示。

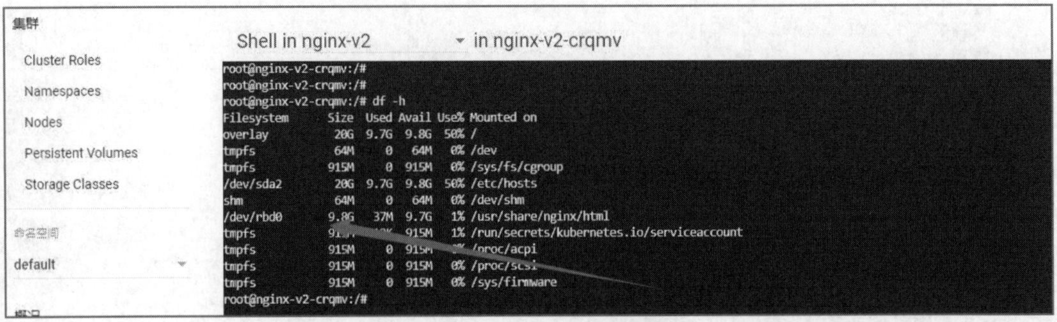

图 8-2　Kubernetes Ceph RBD 模式实战

（3）测试 Kubernetes Pod Ceph 和 Ceph 服务器数据是否一致，如图 8-3 所示。

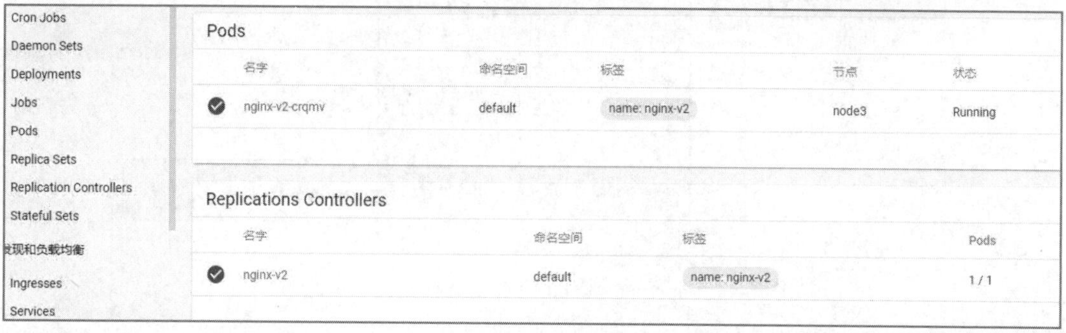

（a）

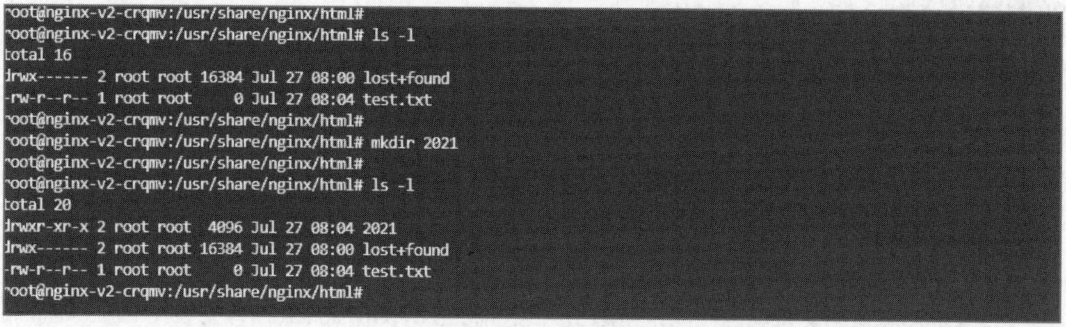

（b）

图 8-3　Kubernetes Ceph RBD 模式实战

8.5　Kubernetes+Ceph RBD 动态存储模式

Kubernetes 使用 Ceph RBD 共享动态存储模式，需要动态创建 PV，是指在现有 PV 不满足 PVC 的请求时，可以使用存储分类（StorageClass），描述具体过程：PV 先创建分类，PVC 请求已创建的某个类（StorageClass）的资源，这样就达到动态配置的效果，即通过一个叫 StorageClass 的对象由存储系统根据 PVC 的要求自动创建。

其中动态方式是通过 StorageClass 来完成的，这是一种新的存储供应方式。动态卷供给能力让管理员不必进行预先创建存储卷，而是随用户需求进行创建，如图 8-4 所示。

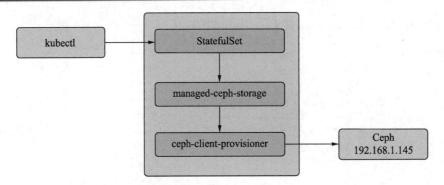

图 8-4　Kubernetes Ceph PVC 动态存储流程示意图

使用 StorageClass 有什么好处呢？除了由存储系统动态创建，节省了管理员的时间外，还可以封装不同类型的存储供 PVC 选用。

在 StorageClass 出现以前，PVC 绑定一个 PV 只能根据两个条件，一个是存储的大小，另一个是访问模式。在 StorageClass 出现后，等于增加了一个绑定维度。

8.6　Ceph RBD 插件配置实战

（1）Ceph 默认不支持动态存储，使用了第三方的 rbd-provisioner 插件安装 rbd-provisioner 插件，GitHub 地址如下：

```
https://kubernetes.io/docs/concepts/storage/storage-classes/
```

（2）下载 rbd-provisioner 和动态 PV 配置文件。

```
cat>external-storage-rbd-provisioner.yaml<<EOF
apiVersion: v1
kind: ServiceAccount
metadata:
  name: rbd-provisioner
  namespace: kube-system
---
kind: ClusterRole
apiVersion: rbac.authorization.Kubernetes.io/v1
metadata:
  name: rbd-provisioner
rules:
  - apiGroups: [""]
```

```yaml
    resources: ["persistentvolumes"]
    verbs: ["get", "list", "watch", "create", "delete"]
  - apiGroups: [""]
    resources: ["persistentvolumeclaims"]
    verbs: ["get", "list","watch", "update"]
  - apiGroups: ["storage.Kubernetes.io"]
    resources: ["storageclasses"]
    verbs: ["get", "list", "watch"]
  - apiGroups: [""]
    resources: ["events"]
    verbs: ["create", "update", "patch"]
  - apiGroups: [""]
    resources: ["endpoints"]
    verbs: ["get", "list", "watch", "create", "update", "patch"]
  - apiGroups: [""]
    resources: ["services"]
    resourceNames: ["kube-dns"]
    verbs: ["list", "get"]
---
kind: ClusterRoleBinding
apiVersion: rbac.authorization.Kubernetes.io/v1
metadata:
  name: rbd-provisioner
subjects:
  - kind: ServiceAccount
    name: rbd-provisioner
    namespace: kube-system
roleRef:
  kind: ClusterRole
  name: rbd-provisioner
  apiGroup: rbac.authorization.Kubernetes.io

---
apiVersion: rbac.authorization.Kubernetes.io/v1
kind: Role
metadata:
  name: rbd-provisioner
  namespace: kube-system
```

```yaml
rules:
- apiGroups: [""]
  resources: ["secrets"]
  verbs: ["get"]
---
apiVersion: rbac.authorization.Kubernetes.io/v1
kind: RoleBinding
metadata:
  name: rbd-provisioner
  namespace: kube-system
roleRef:
  apiGroup: rbac.authorization.Kubernetes.io
  kind: Role
  name: rbd-provisioner
subjects:
- kind: ServiceAccount
  name: rbd-provisioner
  namespace: kube-system

---
apiVersion: apps/v1
kind: Deployment
metadata:
  name: rbd-provisioner
  namespace: kube-system
spec:
  replicas: 1
  selector:
    matchLabels:
      app: rbd-provisioner
  strategy:
    type: Recreate
  template:
    metadata:
      labels:
        app: rbd-provisioner
    spec:
      containers:
```

```yaml
      - name: rbd-provisioner
        image: "quay.io/external_storage/rbd-provisioner:latest"
        env:
        - name: PROVISIONER_NAME
          value: ceph.com/rbd
      serviceAccount: rbd-provisioner
EOF
```

（3）执行以上 YAML 文件，使其生效，并查看其 Pod 信息。

```
kubectl apply -f external-storage-rbd-provisioner.yaml
```

（4）查看 Pod 状态，如图 8-5 所示，操作命令如下：

```
kubectl get pod -n kube-system |grep -aiE provisioner
```

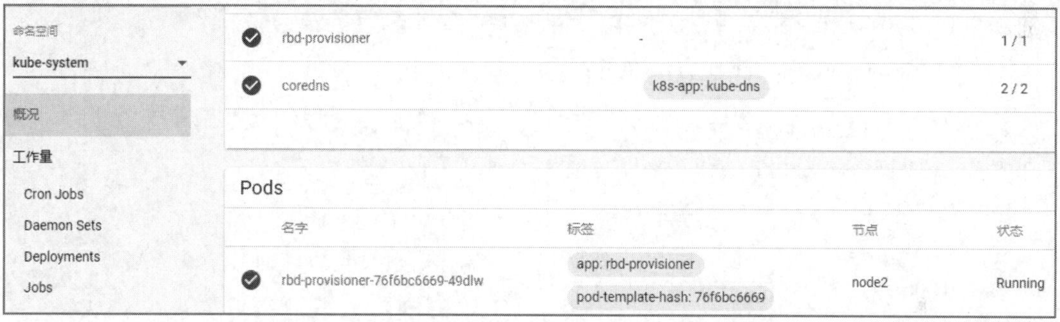

图 8-5　Kubernetes Ceph RBD Provisioner 创建

（5）创建 RBD OSD 资源池，操作指令如下：

```
#Kubernetes 集群中所有节点安装 ceph-common
#后期需要使用 rdb 命令 map 附加 rbd 创建的 image
yum install -y ceph-common
#在 ceph 的 mon 或者 admin 节点创建 osd pool
ceph osd pool create kube 4096
ceph osd pool ls
#在 ceph 的 mon 或者 admin 节点创建 Kubernetes 访问 ceph 的用户
ceph auth get-or-create client.kube mon 'allow r' osd 'allow class-read
object_prefix rbd_children, allow rwx pool=kube' -o ceph.client.kube.keyring
#查看 key 在 ceph 的 mon 或者 admin 节点
ceph auth get-key client.admin
ceph auth get-key client.kube
#创建 admin secret
```

```
#将 CEPH_ADMIN_SECRET 替换为 client.admin 获取到的 key
export CEPH_ADMIN_SECRET='AQCXgP5giXmCARAAzz1mWKqJ+dTFdLArr1Ee+Q=='
kubectl create secret generic ceph-secret --type="kubernetes.io/rbd" \
--from-literal=key=$CEPH_ADMIN_SECRET \
--namespace=kube-system
#在 default 命名空间创建 pvc,用于访问 ceph 的 secret
#将 CEPH_KUBE_SECRET 替换为 client.kube 获取到的 key
export CEPH_KUBE_SECRET='AQAQ/P9gcQSvARAAek8WxixlCnb4vMChm3eLhA=='
kubectl create secret generic ceph-user-secret --type="kubernetes.io/rbd" \
--from-literal=key=$CEPH_KUBE_SECRET \
--namespace=default
#查看 secret
kubectl get secret ceph-user-secret -o yaml
kubectl get secret ceph-secret -n kube-system -o yaml
```

(6) 配置 StorageClass,操作方法如下:

```
#如果使用 kubeadm 创建的集群 provisioner,则使用如下方式
# provisioner: ceph.com/rbd
cat >storageclass-ceph-rdb.yaml<<EOF
kind: StorageClass
apiVersion: storage.Kubernetes.io/v1
metadata:
  name: dynamic-ceph-rdb
provisioner: ceph.com/rbd
# provisioner: kubernetes.io/rbd
parameters:
  monitors: 192.168.1.145:6789
  adminId: admin
  adminSecretName: ceph-secret
  adminSecretNamespace: kube-system
  pool: kube
  userId: kube
  userSecretName: ceph-user-secret
  fsType: ext4
  imageFormat: "2"
  imageFeatures: "layering"
EOF
kubectl apply -f storageclass-ceph-rdb.yaml
```

```
kubectl get sc
```

(7)创建 Ceph RBD PVC 资源索取,操作指令如下:

```
cat>ceph-rdb-pvc.yaml<<EOF
kind: PersistentVolumeClaim
apiVersion: v1
metadata:
  name: ceph-rdb-claim
spec:
  accessModes:
    - ReadWriteOnce
  storageClassName: dynamic-ceph-rdb
  resources:
    requests:
      storage: 2Gi
EOF
kubectl apply -f ceph-rdb-pvc.yaml
```

(8)创建 Nginx pod 容器,使用 Ceph RBD 资源池,如图 8-6 所示,操作指令如下:

```
apiVersion: v1
kind: Pod
metadata:
  name: nginx-pod1
  labels:
    name: nginx-pod1
spec:
  containers:
  - name: nginx-pod1
    image: docker.io/library/nginx:latest
    ports:
    - name: web
      containerPort: 80
    volumeMounts:
    - name: ceph-rdb
      mountPath: /usr/share/nginx/html
  volumes:
  - name: ceph-rdb
    persistentVolumeClaim:
      claimName: ceph-rdb-claim
```

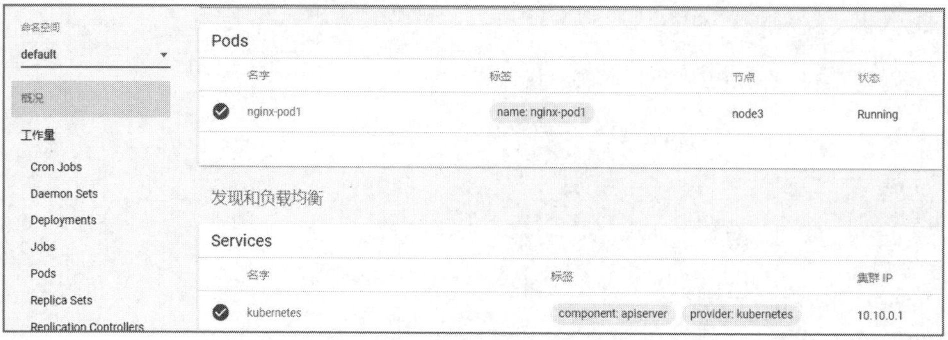

（a）

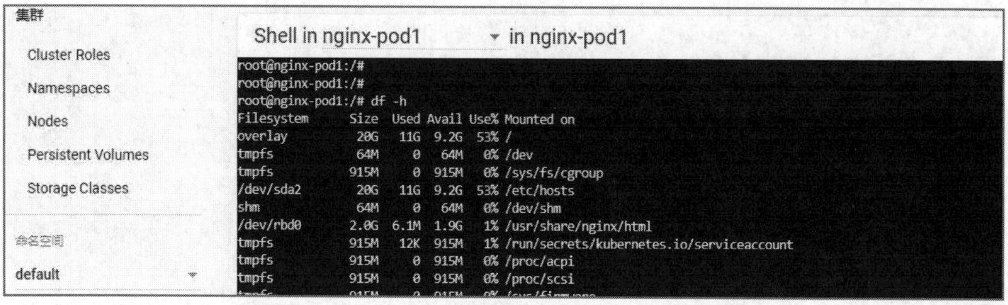

（b）

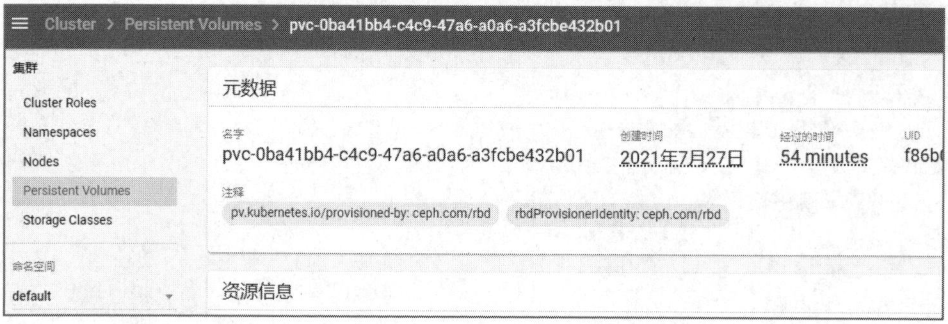

（c）

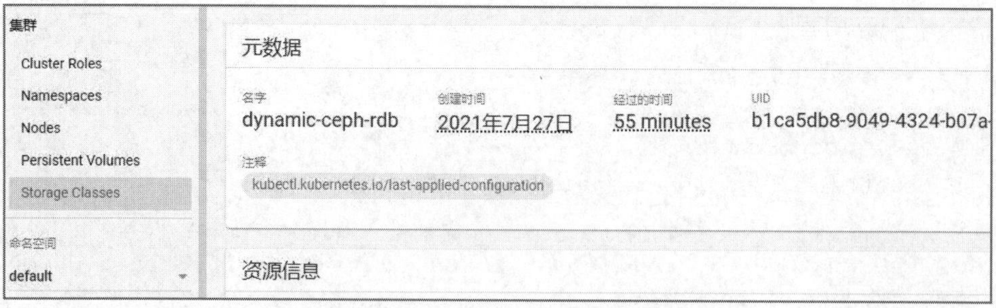

（d）

图 8-6 Kubernetes Ceph RBD 实战

第 9 章 Prometheus 监控 Kubernetes 实战

Prometheus（普罗米修斯）是一套开源的、免费的分布式系统监控报警平台，与 Cacti、Nagios、Zabbix 类似，是企业最常使用的监控系统之一，但是 Prometheus 作为新一代的监控系统，主要应用于云计算方面。

Prometheus 自 2012 成立以来，被许多公司和组织采用，现在是一个独立的开源项目，并独立于任何公司维护。2016 年起，Prometheus 加入云计算基金会作为 Kubernetes 之后的第二托管项目。

9.1 Prometheus 监控优点

Prometheus 相比传统监控系统（Cacti、Nagios、Zabbix），有如下优点。

（1）易管理性：Prometheus 核心部分只有一个单独的二进制文件，可直接在本地工作，不依赖于分布式存储。

（2）业务数据相关性：监控服务的运行状态，基于 Prometheus 丰富的 Client 库，用户可以轻松地在应用程序中添加对 Prometheus 的支持，从而获取服务和应用内部真正的运行状态。

（3）性能高效性：单一 Prometheus 可以处理数以百万的监控指标，每秒处理数十万的数据点。

（4）易于伸缩性：使用功能分区（sharing）+集群（federation）对 Prometheus 进行扩展，形成一个逻辑集群。

（5）良好的可视化：Prometheus 除了自带 Prometheus UI，还提供了一个独立的基于 Ruby On

Rails 的 Dashboard 解决方案 Promdash。另外，最新的 Grafana 可视化工具也提供了完整的 Prometheus 支持，基于 Prometheus 提供的 API 还可以实现自己的监控可视化 UI。

9.2 Prometheus 监控特点

Prometheus 监控特点如下：
（1）由度量名和键值对标识的时间序列数据的多维数据模型。
（2）灵活的查询语言。
（3）不依赖于分布式存储，单服务器节点是自治的。
（4）通过 HTTP 上的拉模型实现时间序列收集。
（5）通过中间网关支持推送时间序列。
（6）通过服务发现或静态配置发现目标。
（7）图形和仪表板支持的多种模式。

9.3 Prometheus 组件实战

Prometheus 生态由多个组件组成，并且这些组件大部分是可选的。

1. Prometheus server

Prometheus server（Prometheus 服务端）是 Prometheus 组件中的核心部分，负责实现对监控数据的获取、存储及查询。Prometheus server 可以通过静态配置管理监控目标，也可以配合使用 service discovery 的方式动态管理监控目标，并从这些监控目标中获取数据。

其次，Prometheus server 需要对采集到的数据进行存储，Prometheus server 本身就是一个实时数据库，将采集到的监控数据按照时间序列的方式存储在本地磁盘中。Prometheus server 对外提供了自定义的 PromQL，实现对数据的查询以及分析。

另外，Prometheus server 的联邦集群能力可以使其从其他 Prometheus server 实例中获取数据。

2. Exporter监控客户端

Exporter 将监控数据采集的端点通过 HTTP 服务的形式暴露给 Prometheus server，Prometheus server 通过访问该 Exporter 提供的 Endpoint（端点），即可以获取需要采集的监控数据。可以将

Exporter 分为两类。

（1）直接采集：这一类 Exporter 直接内置了对 Prometheus 监控的支持，如 cAdvisor、Kubernetes、etcd、Gokit 等，都直接内置了用于向 Prometheus 暴露监控数据的端点。

（2）间接采集：原有监控目标并不直接支持 Prometheus，需要通过 Prometheus 提供的 Client Library 编写该监控目标的监控采集程序，如 MySQL Exporter、JMX Exporter、Consul Exporter 等。

3. Alertmanager报警模块

在 Prometheus server 中支持基于 PromQL 创建告警规则，如果满足 PromQL 定义的规则，则会产生一条告警。在 Alertmanager 从 Prometheus server 端接收到告警后，会进行去除重复数据、分组，并路由到对收的接受方式，发出报警。常见的接收方式有电子邮件、pagerduty、webhook 等。

4. PushGateway网关

Prometheus 数据采集基于 Prometheus server 从 Exporter 拉取数据，因此当网络环境不允许 Prometheus server 和 Exporter 进行通信时，可以使用 PushGateway 进行中转。通过 PushGateway 将内部网络的监控数据主动推送到 Gateway 中，Prometheus server 采用针对 Exporter 同样的方式，将监控数据从 PushGateway 拉取到 Prometheus server。

5. Web UI平台

Prometheus 的 Web 接口可用于简单可视化，以及语句执行或者服务状态监控。

9.4 Prometheus 体系结构

Prometheus 从 jobs 获取度量数据，也可以直接或通过推送网关获取临时 jobs 的度量数据。它在本地存储所有被获取的样本，并在这些数据运行规则中，对现有数据进行聚合和记录新的时间序列或生成警报。

通过 Grafana 或其他 API 消费者，可以可视化地查看收集到的数据。Prometheus 的整体架构和生态组件如图 9-1 所示。

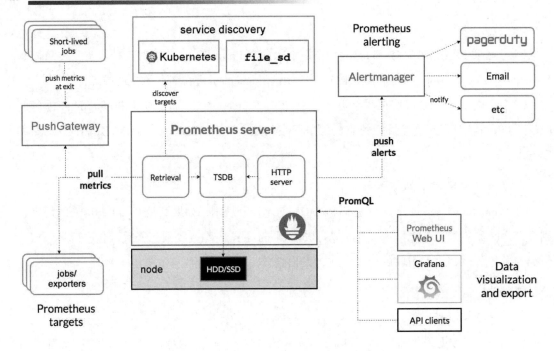

图 9-1 Prometheus 的整体架构和生态组件

9.5 Prometheus 工作流程

（1）Prometheus 服务器定期从配置好的 jobs 或 Exporter 中获取度量数据，或者接收来自推送网关发送过来的度量数据。

（2）Prometheus 服务器在本地存储收集到的度量数据，并对这些数据进行聚合。

（3）运行已定义好的 alert.rules，记录新的时间序列或者向告警管理器推送警报。

（4）告警管理器根据配置文件，对接收到的警报进行处理，并通过 Email、微信、钉钉等途径发出告警。

（5）Grafana 等图形工具获取到监控数据，并以图形化的方式进行展示。

9.6 Prometheus 和 Kubernetes 背景

在 Kubernetes 集群中部署 node-exporter、Prometheus、Grafana，同时使用 Prometheus 对 Kubernetes 整个集群进行监控。实现方法和原理如下：

（1）node-exporter 负责收集节点上的 metrics 监控数据，将数据推送给 Prometheus server 端。

（2）Prometheus server 负责存储这些监控数据。

（3）Grafana 将这些数据通过网页以图形的形式展现给用户。

9.7　Kubernetes 集群部署 node-exporter

（1）在 master 和 node 节点下载 Prometheus 相关镜像，操作指令如下：

```
docker pull prom/node-exporter
docker pull prom/prometheus:v2.26.0
docker pull grafana/grafana
```

（2）基于 DaemonSet 方式部署 node-exporter 组件，每个节点只部署一个 node-exporter 实例，操作指令如下：

```
cat>node-exporter.yaml<<EOF
apiVersion: apps/v1
kind: DaemonSet
metadata:
  name: node-exporter
  namespace: kube-system
  labels:
    Kubernetes-app: node-exporter
spec:
  selector:
    matchLabels:
      Kubernetes-app: node-exporter
  template:
    metadata:
      labels:
        Kubernetes-app: node-exporter
    spec:
      containers:
      - image: prom/node-exporter
        name: node-exporter
        ports:
        - containerPort: 9100
          protocol: TCP
          name: http
---
```

```
apiVersion: v1
kind: Service
metadata:
  labels:
    Kubernetes-app: node-exporter
  name: node-exporter
  namespace: kube-system
spec:
  ports:
  - name: http
    port: 9100
    nodePort: 31672
    protocol: TCP
  type: NodePort
  selector:
    Kubernetes-app: node-exporter
EOF
kubectl apply -f node-exporter.yaml
```

9.8　Kubernetes 集群部署 Prometheus

（1）部署 Prometheus 相关服务组件，可以从京峰官网下载 YAML 配置文件，操作指令如下：

```
mkdir prometheus/
cd prometheus/
for i in alertmanager-configmap.yaml alertmanager-deployment.yaml
alertmanager-pvc.yaml configmap.yaml grafana-deploy.yaml grafana-service.
yaml node-exporter.yaml prometheus.deploy.yml prometheus-rules.yaml
prometheus.svc.yaml rbac-setup.yaml ;do wget -c http://bbs.jingfengjiaoyu.
com/download/docker/prometheus/$i ;done
#批量应用以上 YAML 脚本
for i in alertmanager-configmap.yaml alertmanager-deployment.yaml
alertmanager-pvc.yaml configmap.yaml grafana-deploy.yaml grafana-service.
yaml node-exporter.yaml prometheus.deploy.yml prometheus-rules.yaml
prometheus.svc.yaml rbac-setup.yaml ;do kubectl apply -f $i ;sleep 3 ;done
```

（2）部署 Prometheus 相关服务组件，部署 Rbac 认证文件 rbac-setup.yaml，操作指令如下：

```
apiVersion: rbac.authorization.Kubernetes.io/v1
kind: ClusterRole
metadata:
```

```yaml
  name: prometheus
rules:
- apiGroups: [""]
  resources:
  - nodes
  - nodes/proxy
  - services
  - endpoints
  - pods
  verbs: ["get", "list", "watch"]
- apiGroups:
  - extensions
  resources:
  - ingresses
  verbs: ["get", "list", "watch"]
- nonResourceURLs: ["/metrics"]
  verbs: ["get"]
---
apiVersion: v1
kind: ServiceAccount
metadata:
  name: prometheus
  namespace: kube-system
---
apiVersion: rbac.authorization.Kubernetes.io/v1
kind: ClusterRoleBinding
metadata:
  name: prometheus
roleRef:
  apiGroup: rbac.authorization.Kubernetes.io
  kind: ClusterRole
  name: prometheus
subjects:
- kind: ServiceAccount
  name: prometheus
  namespace: kube-system
```

（3）部署 Prometheus 相关服务组件，部署 Prometheus 主程序，保存为文件 prometheus.deploy.yml，操作指令如下：

```yaml
---
apiVersion: apps/v1
```

```yaml
kind: Deployment
metadata:
  labels:
    name: prometheus-deployment
  name: prometheus
  namespace: kube-system
spec:
  replicas: 1
  selector:
    matchLabels:
      app: prometheus
  template:
    metadata:
      labels:
        app: prometheus
    spec:
      containers:
      - image: prom/prometheus:v2.26.0
        name: prometheus
        command:
        - "/bin/prometheus"
        args:
        - "--config.file=/etc/prometheus/prometheus.yml"
        - "--storage.tsdb.path=/prometheus"
        - "--storage.tsdb.retention=24h"
        ports:
        - containerPort: 9090
          protocol: TCP
        volumeMounts:
        - mountPath: "/prometheus"
          name: data
        - mountPath: "/etc/prometheus"
          name: config-volume
        resources:
          requests:
            cpu: 100m
            memory: 100Mi
          limits:
            cpu: 500m
            memory: 2500Mi
      serviceAccountName: prometheus
```

```yaml
      volumes:
      - name: data
        emptyDir: {}
      - name: config-volume
        configMap:
          name: prometheus-config
```

（4）部署 Prometheus 相关服务组件，部署 Prometheus service，保存为文件 prometheus.svc.yml，操作指令如下：

```yaml
kind: Service
apiVersion: v1
metadata:
  labels:
    app: prometheus
  name: prometheus
  namespace: kube-system
spec:
  type: NodePort
  ports:
  - port: 9090
    targetPort: 9090
    nodePort: 30003
  selector:
    app: prometheus
```

（5）以 ConfigMap 的形式管理 Prometheus 组件的配置文件 configmap.yaml，操作指令如下：

```yaml
apiVersion: v1
kind: ConfigMap
metadata:
  name: prometheus-config
  namespace: kube-system
data:
  prometheus.yml: |
    global:
      scrape_interval:     15s
      evaluation_interval: 15s
    scrape_configs:

    - job_name: 'kubernetes-apiservers'
      kubernetes_sd_configs:
      - role: endpoints
```

```yaml
      scheme: https
      tls_config:
        ca_file: /var/run/secrets/kubernetes.io/serviceaccount/ca.crt
        bearer_token_file: /var/run/secrets/kubernetes.io/serviceaccount/token
      relabel_configs:
      - source_labels: [__meta_kubernetes_namespace, __meta_kubernetes_service_name, __meta_kubernetes_endpoint_port_name]
        action: keep
        regex: default;kubernetes;https

    - job_name: 'kubernetes-nodes'
      kubernetes_sd_configs:
      - role: node
      scheme: https
      tls_config:
        ca_file: /var/run/secrets/kubernetes.io/serviceaccount/ca.crt
        bearer_token_file: /var/run/secrets/kubernetes.io/serviceaccount/token
      relabel_configs:
      - action: labelmap
        regex: __meta_kubernetes_node_label_(.+)
      - target_label: __address__
        replacement: kubernetes.default.svc:443
      - source_labels: [__meta_kubernetes_node_name]
        regex: (.+)
        target_label: __metrics_path__
        replacement: /api/v1/nodes/${1}/proxy/metrics

    - job_name: 'kubernetes-cadvisor'
      kubernetes_sd_configs:
      - role: node
      scheme: https
      tls_config:
        ca_file: /var/run/secrets/kubernetes.io/serviceaccount/ca.crt
        bearer_token_file: /var/run/secrets/kubernetes.io/serviceaccount/token
      relabel_configs:
      - action: labelmap
        regex: __meta_kubernetes_node_label_(.+)
      - target_label: __address__
```

```yaml
        replacement: kubernetes.default.svc:443
      - source_labels: [__meta_kubernetes_node_name]
        regex: (.+)
        target_label: __metrics_path__
        replacement: /api/v1/nodes/${1}/proxy/metrics/cadvisor

    - job_name: 'kubernetes-service-endpoints'
      kubernetes_sd_configs:
      - role: endpoints
      relabel_configs:
      - source_labels: [__meta_kubernetes_service_annotation_prometheus_io_scrape]
        action: keep
        regex: true
      - source_labels: [__meta_kubernetes_service_annotation_prometheus_io_scheme]
        action: replace
        target_label: __scheme__
        regex: (https?)
      - source_labels: [__meta_kubernetes_service_annotation_prometheus_io_path]
        action: replace
        target_label: __metrics_path__
        regex: (.+)
      - source_labels: [__address__, __meta_kubernetes_service_annotation_prometheus_io_port]
        action: replace
        target_label: __address__
        regex: ([^:]+)(?::\d+)?;(\d+)
        replacement: $1:$2
      - action: labelmap
        regex: __meta_kubernetes_service_label_(.+)
      - source_labels: [__meta_kubernetes_namespace]
        action: replace
        target_label: kubernetes_namespace
      - source_labels: [__meta_kubernetes_service_name]
        action: replace
        target_label: kubernetes_name

    - job_name: 'kubernetes-services'
      kubernetes_sd_configs:
```

```
      - role: service
    metrics_path: /probe
    params:
      module: [http_2xx]
    relabel_configs:
    - source_labels: [__meta_kubernetes_service_annotation_prometheus_io_probe]
      action: keep
      regex: true
    - source_labels: [__address__]
      target_label: __param_target
    - target_label: __address__
      replacement: blackbox-exporter.example.com:9115
    - source_labels: [__param_target]
      target_label: instance
    - action: labelmap
      regex: __meta_kubernetes_service_label_(.+)
    - source_labels: [__meta_kubernetes_namespace]
      target_label: kubernetes_namespace
    - source_labels: [__meta_kubernetes_service_name]
      target_label: kubernetes_name

  - job_name: 'kubernetes-ingresses'
    kubernetes_sd_configs:
    - role: ingress
    relabel_configs:
    - source_labels: [__meta_kubernetes_ingress_annotation_prometheus_io_probe]
      action: keep
      regex: true
    - source_labels: [__meta_kubernetes_ingress_scheme,__address__,__meta_kubernetes_ingress_path]
      regex: (.+);(.+);(.+)
      replacement: ${1}://${2}${3}
      target_label: __param_target
    - target_label: __address__
      replacement: blackbox-exporter.example.com:9115
    - source_labels: [__param_target]
      target_label: instance
    - action: labelmap
      regex: __meta_kubernetes_ingress_label_(.+)
```

```yaml
        - source_labels: [__meta_kubernetes_namespace]
          target_label: kubernetes_namespace
        - source_labels: [__meta_kubernetes_ingress_name]
          target_label: kubernetes_name

      - job_name: 'kubernetes-pods'
        kubernetes_sd_configs:
        - role: pod
        relabel_configs:
        - source_labels: [__meta_kubernetes_pod_annotation_prometheus_io_scrape]
          action: keep
          regex: true
        - source_labels: [__meta_kubernetes_pod_annotation_prometheus_io_path]
          action: replace
          target_label: __metrics_path__
          regex: (.+)
        - source_labels: [__address__, __meta_kubernetes_pod_annotation_prometheus_io_port]
          action: replace
          regex: ([^:]+)(?::\d+)?;(\d+)
          replacement: $1:$2
          target_label: __address__
        - action: labelmap
          regex: __meta_kubernetes_pod_label_(.+)
        - source_labels: [__meta_kubernetes_namespace]
          action: replace
          target_label: kubernetes_namespace
        - source_labels: [__meta_kubernetes_pod_name]
          action: replace
          target_label: kubernetes_pod_name
```

9.9 Kubernetes 集群部署 Grafana

（1）部署 Grafana Web 图形展示界面，操作指令如下：

```
cat>grafana-deploy.yaml<<EOF
apiVersion: apps/v1
kind: Deployment
```

```yaml
metadata:
  name: grafana-core
  namespace: kube-system
  labels:
    app: grafana
    component: core
spec:
  replicas: 1
  selector:
    matchLabels:
      app: grafana
  template:
    metadata:
      labels:
        app: grafana
        component: core
    spec:
      containers:
      - image: grafana/grafana:4.2.0
        name: grafana-core
        imagePullPolicy: IfNotPresent
        # env:
        resources:
        # keep request = limit to keep this container in guaranteed class
          limits:
            cpu: 100m
            memory: 500Mi
          requests:
            cpu: 100m
            memory: 500Mi
        env:
          # The following env variables set up basic auth twith the default admin user and admin password
          - name: GF_AUTH_BASIC_ENABLED
            value: "true"
          - name: GF_AUTH_ANONYMOUS_ENABLED
            value: "false"
          # - name: GF_AUTH_ANONYMOUS_ORG_ROLE
          #   value: Admin
          # does not really work, because of template variables in exported dashboards
```

```yaml
        # - name: GF_DASHBOARDS_JSON_ENABLED
        #   value: "true"
        readinessProbe:
          httpGet:
            path: /login
            port: 3000
        # initialDelaySeconds: 30
        # timeoutSeconds: 1
        volumeMounts:
        - name: grafana-persistent-storage
          mountPath: /var
      volumes:
      - name: grafana-persistent-storage
        emptyDir: {}
EOF
kubectl apply -f grafana-deploy.yaml
```

（2）部署 Grafana service 和对外暴露 node port 端口，操作指令如下：

```yaml
cat>grafana-service.yaml<<EOF
apiVersion: v1
kind: Service
metadata:
  name: grafana
  namespace: kube-system
  labels:
    app: grafana
    component: core
spec:
  type: NodePort
  ports:
    - port: 3000
  selector:
    app: grafana
    component: core
EOF
kubectl apply -f grafana-service.yaml
```

9.10　Kubernetes 配置和整合 Prometheus

（1）查看 node-exporter Pod 是否允许，通过浏览器访问任意 node 节点 IP 地址（http://192.168.1.146:31672/metrics），如图 9-2 所示。

图 9-2 Prometheus 客户端 Exporter

（2）Prometheus 对应的 NodePort 端口为 30003，通过访问 node 节点 IP 地址（http://192.168.1.146:30003/targets），可以看到 Prometheus 已经成功连接上了 Kubernetes 的 API server（状态全部为 UP），如图 9-3 所示。

（a）

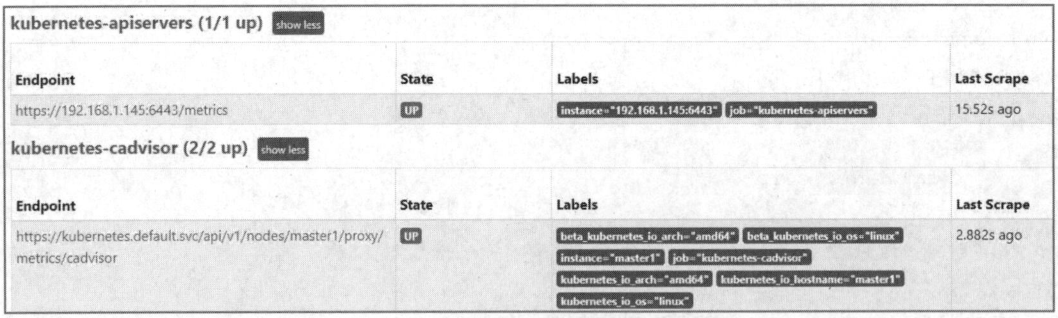

（b）

图 9-3 Prometheus 平台实战操作

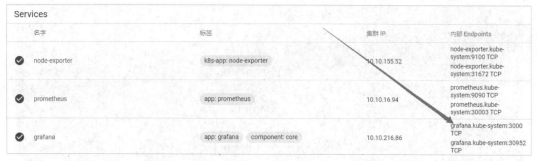

（c）

图 9-3 （续）

（3）通过端口进行 Grafana 访问，默认用户名、密码均为 admin，通过浏览器访问 node 节点 IP 地址（http://192.168.1.146:15493/?orgId=1），添加数据源，如图 9-4 所示。

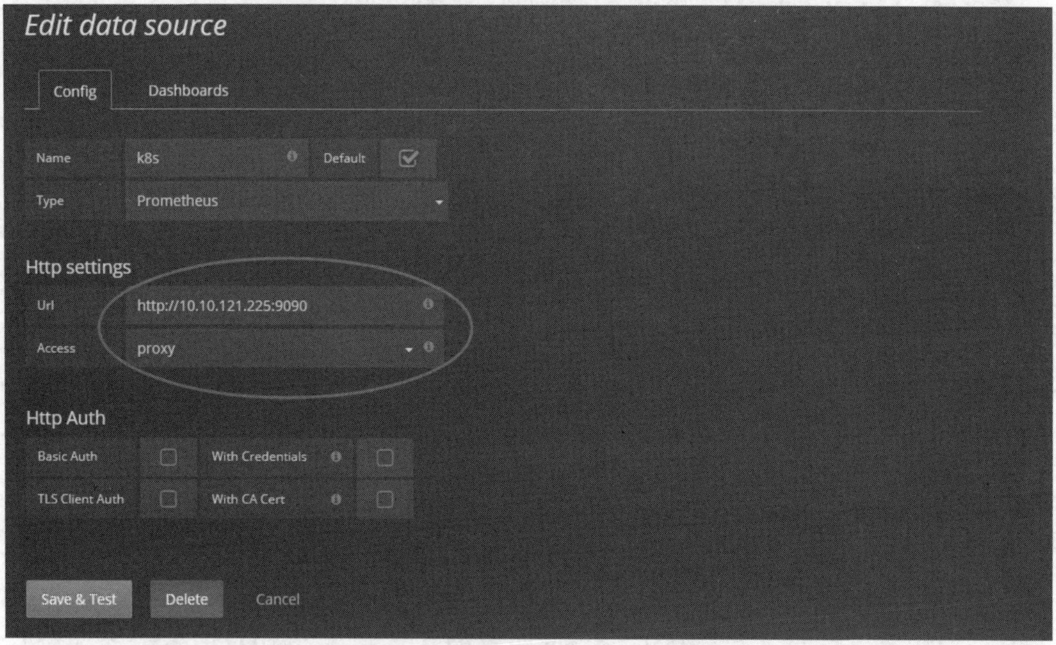

图 9-4 Grafana+Prometheus 整合操作

（4）导入 Grafana 模板，填写 ID（315|3119）即可，也可以从 Grafana 官网选择模板：https://grafana.com/grafana/dashboards，如图 9-5 所示。

（5）最后查看 Prometheus 监控 Kubernetes node 数据展示信息，如图 9-6 所示。

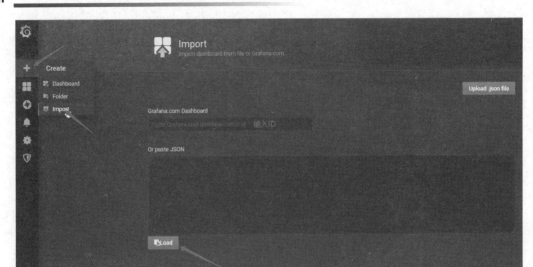

图 9-5 Grafana 平台导入模板操作

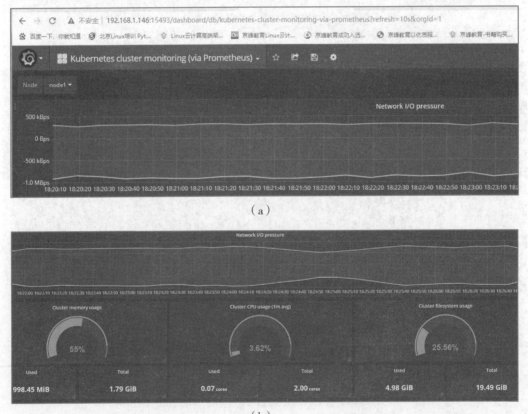

图 9-6 Prometheus 监控 Kubernetes 展示界面

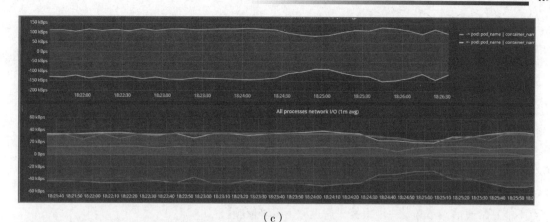

(c)

(d)

图 9-6 （续）

9.11 Kubernetes+Prometheus 报警设置

Prometheus 触发一条告警的过程：Prometheus→触发阈值→超出持续时间→Alertmanager→分组|抑制|静默→媒体类型→邮件、钉钉、微信等。

1. 分组（group）

将类似性质的警报合并为单个通知。例如，Web 服务是一组，CPU 是一组，不用发多个 CPU 超出范围的报警，只发单个 CPU 组的报警即可。

2. 静默（silences）

是一种简单的特定时间静音的机制。例如，服务器要升级维护，可以先设置这个时间段告警静默，关闭其间的报警，否则在代码升级时，会触发一些报警。

3. 抑制（inhibition）

当警报发出后，停止重复发送由此警报引发的其他警报，即合并一个故障引起的多个报警事件，可以消除冗余告警。例如，一个交换机上有 3 个节点，但由于交换机故障，导致 3 个节点网络不通，所以只发交换机的报警，而不发 3 个节点的网络报警，因为是交换机故障导致节点不能正常通信，不必发一堆报警，这也称为报警依赖。

9.12　Kubernetes Alertmanager 实战

（1）部署 Alertmanager，通过 Kubernetes+Prometheus 整合方式部署，需要部署文件 alertmanager-configmap.yaml，代码如下：

```yaml
apiVersion: v1
kind: ConfigMap
metadata:
  #配置文件名称
  name: alertmanager
  namespace: kube-system
  labels:
    kubernetes.io/cluster-service: "true"
    addonmanager.kubernetes.io/mode: EnsureExists
data:
  alertmanager.yml: |
    global:
      resolve_timeout: 5m
      #告警自定义邮件
      smtp_smarthost: 'smtp.163.com:25'
      smtp_from: 'wgkgood@163.com'
      smtp_auth_username: 'wgkgood@163.com'
      smtp_auth_password: 'FNGDCYDBFSXRLUUC'

    receivers:
    - name: default-receiver
      email_configs:
```

```
      - to: "wgkgood@163.com"

    route:
      group_interval: 1m
      group_wait: 10s
      receiver: default-receiver
      repeat_interval: 1m
```

（2）部署文件 alertmanager-deployment.yaml，代码如下：

```
apiVersion: apps/v1
kind: Deployment
metadata:
  labels:
    name: alertmanager
  name: alertmanager
  namespace: kube-system
spec:
  replicas: 1
  selector:
    matchLabels:
      app: alertmanager
  template:
    metadata:
      labels:
        app: alertmanager
    spec:
      containers:
      - image: prom/alertmanager:v0.16.1
        name: alertmanager
        ports:
        - containerPort: 9093
          protocol: TCP
        volumeMounts:
        - mountPath: "/alertmanager"
          name: data
        - mountPath: "/etc/alertmanager"
          name: config-volume
        resources:
          requests:
            cpu: 50m
            memory: 50Mi
```

```yaml
            limits:
              cpu: 200m
              memory: 200Mi
        volumes:
        - name: data
          emptyDir: {}
        - name: config-volume
          configMap:
            name: alertmanager
---
apiVersion: v1
kind: Service
metadata:
  labels:
    app: alertmanager
  annotations:
    prometheus.io/scrape: 'true'
  name: alertmanager
  namespace: kube-system
spec:
  type: NodePort
  ports:
  - port: 9093
    targetPort: 9093
    nodePort: 31113
  selector:
    app: alertmanager
```

（3）部署文件 alertmanager-pvc.yaml，代码如下：

```yaml
apiVersion: v1
kind: PersistentVolumeClaim
metadata:
  name: alertmanager
  namespace: kube-system
  labels:
    kubernetes.io/cluster-service: "true"
    addonmanager.kubernetes.io/mode: EnsureExists
spec:
  #使用自己的动态 PV
  storageClassName: managed-nfs-storage
  accessModes:
```

```
      - ReadWriteOnce
  resources:
    requests:
      storage: "2Gi"
```

（4）部署文件 prometheus-rules.yaml，代码如下：

```
apiVersion: v1
kind: ConfigMap
metadata:
  name: prometheus-rules
  namespace: kube-system
data:
  #通用角色
  general.rules: |
    groups:
    - name: general.rules
      rules:
      - alert: InstanceDown
        expr: up == 0
        for: 1m
        labels:
          severity: error
        annotations:
          summary: "Instance {{ $labels.instance }} 停止工作"
          description: "{{ $labels.instance }} job {{ $labels.job }} 已经停止 5 分钟以上."
  #node 对所有资源的监控
  node.rules: |
    groups:
    - name: node.rules
      rules:
      - alert: NodeFilesystemUsage
        expr: 100 - (node_filesystem_free_bytes{fstype=~"ext4|xfs"} / node_filesystem_size_bytes{fstype=~"ext4|xfs"} * 100) > 80
        for: 1m
        labels:
          severity: warning
        annotations:
          summary: "Instance {{ $labels.instance }} : {{ $labels.mountpoint }} 分区使用率过高"
          description: "{{ $labels.instance }}: {{ $labels.mountpoint }} 分
```

区使用大于 80% (当前值: {{ $value }})"

```
  - alert: NodeMemoryUsage
    expr: 100 - (node_memory_MemFree_bytes+node_memory_Cached_bytes+node_memory_Buffers_bytes) / node_memory_MemTotal_bytes * 100 > 80
    for: 1m
    labels:
      severity: warning
    annotations:
      summary: "Instance {{ $labels.instance }} 内存使用率过高"
      description: "{{ $labels.instance }}内存使用大于80% (当前值: {{ $value }})"

  - alert: NodeCPUUsage
    expr: 100 - (avg(irate(node_cpu_seconds_total{mode="idle"}[5m])) by (instance) * 100) > 60
    for: 1m
    labels:
      severity: warning
    annotations:
      summary: "Instance {{ $labels.instance }} CPU 使用率过高"
      description: "{{ $labels.instance }}CPU 使用大于60% (当前值: {{ $value }})"
```

9.13 Alertmanager 实战部署

（1）部署 Alertmanager，单独部署在一台服务器上，默认监听端口是 9093，部署方法如下：

```
cd /usr/local/src
tar xvf alertmanager-0.20.0.linux-amd64.tar.gz
ln -s /usr/local/src/alertmanager-0.20.0.linux-amd64 /usr/local/alertmanager
cd /usr/local/alertmanager
```

（2）修改 Alertmanager 配置文件 vim alertmanager.yml，配置文件代码如下：

```
global:
  resolve_timeout: 5m                           #超时时间
  smtp_smarthost: 'smtp.163.com'                #smtp 服务器地址
  smtp_from: 'wgkgood@163.com'                  #发件人
  smtp_auth_username: 'wgkgood'                 #登录认证的用户名
  smtp_auth_password: 'jfedu666'                #登录的授权码
  smtp_hello: '@163.com'
```

```
  smtp_require_tls: false        #是否使用tls
  route:                         #route用来设置报警的分发策略,由谁去发
    group_by: ['alertname']      #采用哪个标签作为分组依据
    group_wait: 10s              #组告警等待时间,也就是告警产生后等待10s,一个组内的
                                 #告警10s后一起发送出去
    group_interval: 10s          #两组告警的间隔时间
    repeat_interval: 2m          #重复告警的间隔时间,减少相同邮件的发送频率
    receiver: 'web.hook'         #设置接收人;真正发送邮件不是由route完成,而是由
                                 #receiver发送邮件receivers:- name: 'web.hook'
  #webhook_configs:              #调用指定的API把邮件发送出去
  #- url: 'http://127.0.0.1:5001/'
    email_configs:               #通过邮件的方式发送
  - to: 'wgkgood@163.com'        #接收人
    - source_match:              #源匹配级别;以下设置的级别的报警不会发送
        severity: 'critical'
      target_match:              #目标匹配级别
        severity: 'warning'
      equal: ['alertname', 'dev', 'instance']
```

(3)将Alertmanager设置为系统服务,操作方法和指令如下:

```
vim /etc/systemd/system/alertmanager.service
[Unit]Description=Prometheus AlertManagerDocumentation=https://prometheus.
io/docs/introduction/overview/After=network.target
[Service]Restart=on-failureExecStart=/usr/local/alertmanager/alertmanager
--config.file=/usr/local/alertmanager/alertmanager.yml
[Install]WantedBy=multi-user.target
systemctl start alertmanager.service
systemctl enable alertmanager.service
```

(4)Prometheus添加规则,绑定Alertmanager服务,操作指令如下:

```
vim /usr/local/prometheus/prometheus.yml
alerting:                        #当触发告警时,把告警发送给下面所配置的服务
  alertmanagers:                 #当出告警时,把告警通知发送给alertmanager
  - static_configs:
    - targets:
      - 192.168.3.146:9093       #指定alertmanager地址及端口
  rule_files:                    #报警规则文件
  - "/usr/local/prometheus/rules.yml"   #指定报警的rules.yml文件所在路径
```

(5)创建Prometheus报警规则,新建rules.yml文件,内容如下:

```
vim /usr/local/prometheus/rules.yml
groups:
  - name: linux_pod.rules        #指定名称
    rules:
    - alert: Pod_all_cpu_usage   #相当于 Zabbix 中的监控项；也是邮件的标题
      expr: (sum by(name)(rate(container_cpu_usage_seconds_total{image!=""}
[5m]))*100) > 75          #PromQL 语句查询到所有 Pod 的 CPU 利用率与后面的值作对比，查
                          #询到的是浮点数，需要乘以 100，转换成整数
      for: 5m                    #每 5min 获取一次 Pod 的 CPU 利用率
      labels:
        severity: critical
        service: pods
      annotations:               #此为当前所有容器的 CPU 利用率
        description: 容器 {{ $labels.name }} CPU 资源利用率大于 75%, (current
value is {{ $value }})    #报警的描述信息内容
        summary: Dev CPU 负载告警
    - alert: Pod_all_memory_usage
      expr: sort_desc(avg by(name)(irate(container_memory_usage_bytes
{name!=""} [5m]))*100) > 1024^3*2      #通过 PromQL 语句获取到所有 Pod 中的内存
                                       #利用率，将后面的单位 GB 转换成字节
      for: 10m
      labels:
        severity: critical
      annotations:
        description: 容器 {{ $labels.name }} Memory 资源利用率大于 2GB(当前
已用内存是: {{ $value }})
        summary: Dev Memory 负载告警
    - alert: Pod_all_network_receive_usage
      expr: sum by (name)(irate(container_network_receive_bytes_total
{container_name="POD"}[1m])) > 1024*1024*50
      for: 10m     #因为获取的所有 Pod 网络利用率是字节，所以把后面对比的 MB 转换成字节
      labels:
        severity: critical
      annotations:
        description: 容器 {{ $labels.name }} network_receive 资源利用率大于
50M , (current value is {{ $value }})
```

以上为 PromQL 语句通过 Grafana 找到相应的监控项，单击 edit 找到相应的 PromQL 语句即可。

（6）访问 Prometheus 的 Web 界面，以确认规则是否构建成功，如图 9-7 所示。

第 9 章 Prometheus 监控 Kubernetes 实战

(a)

```
node.rules

Rule

alert: NodeFilesystemUsage
expr: 100
    - (node_filesystem_free_bytes{fstype=~"ext4|xfs"} / node_filesystem_size_bytes{fstype=~"ext4|xfs"}
    * 100) > 80
for: 1m
labels:
    severity: warning
annotations:
    description: '{{ $labels.instance }}: {{ $labels.mountpoint }} 分区使用大于80% (当前值: {{
      $value }})'
    summary: 'Instance {{ $labels.instance }} : {{ $labels.mountpoint }} 分区使用率过高'
```

(b)

图 9-7 Prometheus 规则的定义

（7）检查 rule.yml 文件语法是否正确，如图 9-8 所示，操作指令如下：

```
promtool check rules rules.yml
```

```
general.rules  node.rules
/etc/prometheus/rules $
/etc/prometheus/rules $ promtool check rules node.rules
Checking node.rules
  SUCCESS: 3 rules found

/etc/prometheus/rules $ promtool check rules general.rules
Checking general.rules
  SUCCESS: 1 rules found

/etc/prometheus/rules $
/etc/prometheus/rules $ promtool check rules node.rules
Checking node.rules
  SUCCESS: 3 rules found
```

图 9-8 Prometheus 规则的检测

（8）列出当前 Alertmanager 服务器的所有告警，如图 9-9 所示，操作指令如下：

```
amtool alert --alertmanager.url=http://10.244.1.14:9093/
```

```
[root@node1 ~]# docker ps |grep alert
db750208a2fa   02e0d8e930da                                              "/bin/ale
manager-7d4d4dbd54-4b7mm_kube-system_e26c05ba-aca2-4b83-a6db-3ebe01a303f3_0
9ce7bb5e6623   registry.aliyuncs.com/google_containers/pause:3.2    "/pause"
d4d4dbd54-4b7mm_kube-system_e26c05ba-aca2-4b83-a6db-3ebe01a303f3_0
[root@node1 ~]#
[root@node1 ~]# docker exec -it db750208a2fa sh
/alertmanager $
/alertmanager $
/alertmanager $ ls
/alertmanager $ amtool alert --alertmanager.url=http://10.244.1.14:9093/
Alertname  Starts At  Summary
/alertmanager $
/alertmanager $
/alertmanager $
```

图 9-9　Prometheus Alertmanager 报警列表 1

（9）增加一台 node 节点，增加成功后将其关机，模拟宕机，测试邮件报警实战，如图 9-10 所示。

kubernetes-nodes (2/3 up) show less

Endpoint	State	Labels
https://kubernetes.default.svc:443/api/v1/nodes/master1/proxy/metrics	UP	beta_kubernetes_io_arch="amd64" beta_kubernetes_io_os="linux" instance="master1" job="kubernetes-nodes" kubernetes_io_arch="amd64" kubernetes_io_hostname="master1" kubernetes_io_os="linux"
https://kubernetes.default.svc:443/api/v1/nodes/node1/proxy/metrics	UP	beta_kubernetes_io_arch="amd64" beta_kubernetes_io_os="linux" instance="node1" job="kubernetes-nodes" kubernetes_io_arch="amd64" kubernetes_io_hostname="node1" kubernetes_io_os="linux"
https://kubernetes.default.svc:443/api/v1/nodes/node2/proxy/metrics	DOWN	beta_kubernetes_io_arch="amd64" beta_kubernetes_io_os="linux" instance="node2" job="kubernetes-nodes"

（a）

```
/alertmanager $ ls
/alertmanager $ amtool alert --alertmanager.url=http://10.244.1.14:9093/
Alertname  Starts At  Summary
/alertmanager $
/alertmanager $
/alertmanager $
/alertmanager $
/alertmanager $ amtool alert --alertmanager.url=http://10.244.1.14:9093/
Alertname     Starts At                Summary
InstanceDown  2021-05-14 09:52:28 UTC  Instance node2 停止工作
InstanceDown  2021-05-14 09:52:28 UTC  Instance node2 停止工作
/alertmanager $
/alertmanager $
```

（b）

图 9-10　Prometheus Alertmanager 报警列表 2

（10）访问 Alertmanager Web 界面，查看报警界面，如图 9-11 所示。

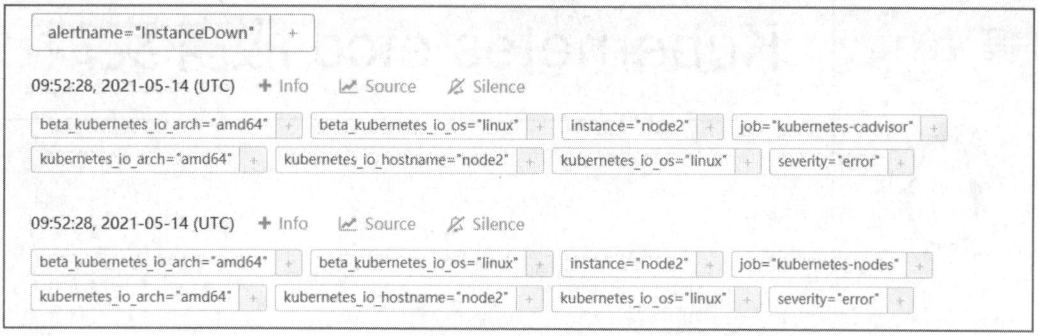

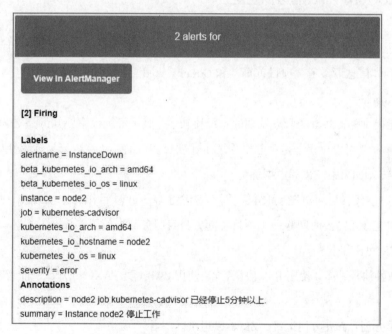

图 9-11　Prometheus Alertmanager 报警界面

（11）登录 163 邮件服务器，查看是否收到 node2 报警信息，如图 9-12 所示。

图 9-12　Prometheus Alertmanager 报警信息

第 10 章 Kubernetes etcd 服务实战

10.1 etcd 和 ZK 服务概念

etcd 是用于共享配置和服务发现的分布式、一致性的 KV 存储系统。该项目目前最新稳定版本为 2.3.0，具体信息请参考"项目首页"和 Github。etcd 是 CoreOS 公司发起的一个开源项目，授权协议为 Apache。

市场上能提供配置共享和服务发现的系统比较多，其中最为大家熟知的是 ZooKeeper（以下简称 ZK），而 etcd 可以算得上是后起之秀了。在项目实现的一致性协议易理解性、运维、安全等多个维度上，etcd 相比 ZK 都占据优势。

ZK 作为典型代表与 etcd 进行比较，而不考虑将 Consul 项目作为比较对象，原因是 Consul 的可靠性和稳定性还需要时间验证（项目发起方自身服务都未使用 Consul）。

etcd 和 ZK 对比如下。

（1）一致性协议：etcd 使用 Raft 协议，ZK 使用 ZAB（类 PAXOS 协议），前者容易理解，方便工程实现。

（2）运维方面：etcd 方便运维，ZK 难以运维。

（3）项目活跃度：etcd 社区与开发活跃，ZK 已经快被用户弃用了。

（4）API：etcd 提供 HTTP+JSON、gRPC 接口，跨平台跨语言，ZK 需要使用其客户端。

（5）访问安全方面：etcd 支持 HTTPS 访问，ZK 在这方面缺失。

10.2 etcd 的使用场景

和 ZK 类似，etcd 有很多使用场景，包括配置管理、服务注册于发现、选主、应用调度、分布式队列、分布式锁。

10.3 etcd 读写性能

按照官网给出的 Benchmark，在 2 个 CPU、1.8GB 内存、SSD 磁盘这样的配置下，单节点的写性能可以达到 16000QPS，而先写后读也能达到 12000QPS。这个性能是相当可观的。

10.4 etcd 工作原理

etcd 使用 Raft 协议维护集群内各个节点状态的一致性。简单地说，etcd 集群是一个分布式系统，由多个节点相互通信构成整体对外服务，每个节点都存储了完整的数据，并通过 Raft 协议保证每个节点维护的数据是一致的，如图 10-1 所示。

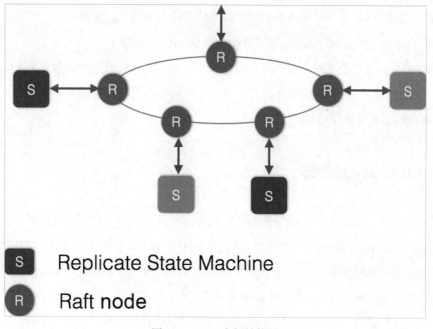

图 10-1　etcd 内部结构图

每个 etcd 节点都维护了一个状态机，且任意时刻至多存在一个有效的主节点。主节点用于处理所有来自客户端的写操作，通过 Raft 协议保证写操作对状态机的改动会可靠地同步到其他节点。

etcd 工作原理核心部分在于 Raft 协议。本节接下来将简要介绍 Raft 协议，具体细节请参考相关论文。

Raft 协议主要分为 3 部分：etcd 选主、etcd 日志复制、etcd 安全性。

10.5 etcd 选主

Raft 协议是用于维护一组服务节点数据一致性的协议。这一组服务节点构成一个集群，且有一个主节点对外提供服务。当集群初始化，或者主节点崩溃后，将面临选主问题。集群中的每个节点在任意时刻处于 Leader、Follower、Candidate 这三个角色之一。选举特点如下：

（1）在集群初始化时，每个节点都是 Follower 角色。

（2）集群中至多存在 1 个有效的主节点，通过心跳与其他节点同步数据。

（3）当 Follower 在一定时间内没有收到来自主节点的心跳时，会将自己角色改变为 Candidate，并发起一次选主投票；当收到包括自己在内超过半数节点赞成后，选举成功；当收到票数不足半数选举失败，或者选举超时。若本轮未选出主节点，将进行下一轮选举（出现这种情况，是由于多个节点同时选举，所有节点均为获得过半选票）。

（4）Candidate 节点收到来自主节点的信息后，会立即终止选举过程，进入 Follower 角色。

（5）为了避免陷入选主失败循环，每个节点未收到心跳发起选举的时间是一定范围内的随机值，这样能够避免 2 个节点同时发起选主。

10.6 etcd 日志复制

所谓日志复制，是指主节点将每次操作形成日志条目，并持久化到本地磁盘，然后通过网络 I/O 发送给其他节点。其他节点根据日志的逻辑时钟（TERM）和日志编号（INDEX）判断是否将该日志记录持久化到本地。当主节点收到包括自己在内超过半数节点成功返回时，认为该日志是可提交的（committed），并将日志输入到状态机，将结果返回客户端。

这里需要注意的是，每次选主都会形成一个唯一的 TERM 编号，相当于逻辑时钟。每一条日志都有全局唯一的编号，如图 10-2 所示。

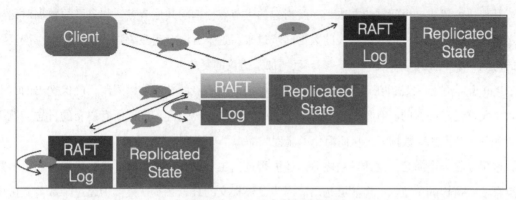

图 10-2　etcd 日志复制结构图

主节点通过网络 I/O 向其他节点追加日志。某节点收到日志追加的消息，将首先判断该日志的 TERM 是否过期，以及该日志条目的 INDEX 是否比当前和提交的日志 INDEX 更早。若已过期，或者比提交的日志更早，则拒绝追加，并返回该节点当前已提交的日志编号；否则，将进行日志追加，并返回成功信息。

当主节点收到其他节点关于日志追加的回复后，若发现被拒绝，则根据该节点返回的已提交日志编号。

主节点向其他节点同步日志，还做了拥塞控制。具体地说，主节点发现日志复制的目标节点拒绝了某次日志追加消息，将进入日志探测阶段，逐条发送日志，直到目标节点接受日志，然后进入快速复制阶段，可进行批量日志追加。

按照日志复制的逻辑可以看到，集群中慢节点不影响整个集群的性能。另外一个特点是，数据只从主节点复制到 Follower 节点，这样大大简化了逻辑流程。

10.7　etcd 安全性

到目前为止，选主以及日志复制并不能保证节点间数据一致。试想，当某个节点挂掉了一段时间后再次重启，并当选为主节点，而在其挂掉这段时间内，集群若有超过半数节点存活，集群会正常工作，那么会有日志提交。因为这些提交的日志无法传递给挂掉的节点，所以当挂掉的节点再次当选主节点时，它将缺失部分已提交的日志。

在这样的场景下，按 Raft 协议，它将自己的日志复制给其他节点，会覆盖集群已经提交的日志。这显然是不可接受的。

其他协议解决这个问题的办法是，新当选的主节点会询问其他节点，和自己的数据对比，确定集群已提交数据，然后将缺失的数据同步过来。这个方案有明显缺陷，增加了集群恢复服务的时间（集群在选举阶段不可服务），且增加了协议的复杂度。

Raft 解决的办法是，在选主逻辑中，对能够成为主节点的节点加以限制，确保选出的节点包含了集群已经提交的所有日志。如果新选出的主节点已经包含了集群所有提交的日志，就不需要和其他节点比对数据了，从而简化了流程，缩短了集群恢复服务的时间。

这里存在一个问题，这样限制之后，还能否选出主节点呢？答案是：只要仍然有超过半数节点存活，这样的主节点一定能够选出。因为已经提交的日志必然被集群中超过半数节点持久化，显然前一个主节点提交的最后一条日志也被集群中大部分节点持久化。

当主节点挂掉后，集群中仍有大部分节点存活，那么这些存活的节点中一定存在一个节点包含了已经提交的日志。

10.8 etcd 使用案例

据公开资料显示，至少有 CoreOS、Google Kubernetes、Cloud Foundry，以及在 Github 上超过 500 个项目在使用 etcd。

10.9 etcd 接口使用

etcd 主要的通信协议是 HTTP，在最新版本中支持 Google gRPC 方式访问。具体支持接口情况如下：

（1）etcd 是一个高可靠的 KV 存储系统，支持 PUT/GET/DELETE 接口。

（2）为了支持服务注册与发现，支持 watch 接口（通过 http long poll 实现）。

（3）支持 key 持有 TTL 属性。

（4）支持 CAS（compare and swap）操作。

（5）支持多 key 的事务操作。

（6）支持目录操作。

第 11 章　Kubernetes+HAProxy 高可用集群

11.1　Kubernetes 高可用集群概念

Kubernetes Apiserver 提供了 Kubernetes 各类资源对象（Pod、RC、service 等）的增删改查及 watch 等 HTTP Rest 接口，是整个系统的数据总线和数据中心。

Kubernetes API server 的功能如下：

（1）提供了集群管理的 Rest API 接口（包括认证授权、数据校验以及集群状态变更）。

（2）提供其他模块之间的数据交互和通信的枢纽（其他模块通过 API server 查询或修改数据，只有 API server 才直接操作 etcd）。

（3）是资源配额控制的入口。

（4）拥有完备的集群安全机制。

API server 是用户和 Kubernetes 集群交互的入口，封装了核心对象的增删改查操作，提供了 RESTFul 风格的 API 接口，通过 etcd 实现持久化并维护对象的一致性。在整个 Kubernetes 集群中，API server 服务至关重要，一旦宕机，整个 Kubernetes 平台将无法使用，所以保障企业高可用是运维必备的工作。

11.2　Kubernetes 高可用工作原理

运行 Keepalived 和 HAProxy 的节点称为 LB（Load balancer, 负载均衡）节点。因为 Keepalived 是一主多备运行模式，所以至少有 2 个 LB 节点。此处采用 master 节点的 2 台机器，HAProxy 监听的端口（8443）需要与 kube-apiserver 的端口 6443 不同，以避免冲突。

下面是基于 Keepalived 和 HAProxy 实现 kube-apiserver 高可用的步骤。

（1）Keepalived 提供 kube-apiserver 对外服务的 VIP。

（2）HAProxy 监听 VIP，后端连接所有 kube-apiserver 实例，提供健康检查和负载均衡功能。

Keepalived 在运行过程中周期检查本机的 HAProxy 进程状态，如果检测到 HAProxy 进程异常，则触发重新选主的过程，VIP 将飘移到新选出来的主节点，从而实现 VIP 的高可用。

最终实现所有组件（如 kubectl、kube-controller-manager、kube-scheduler、kube-proxy 等）都通过 VIP+HAProxy 监听的 8443 端口访问 kube-apiserver 服务，如图 11-1 所示。

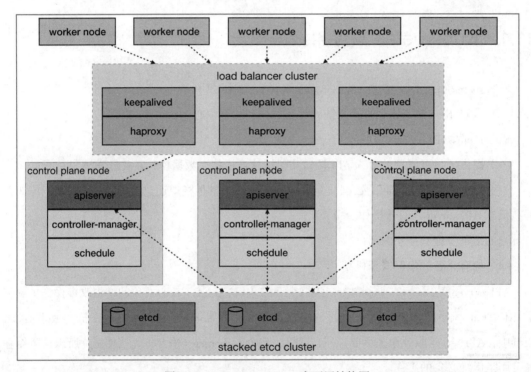

图 11-1　Kubernetes master 高可用结构图

11.3　HAProxy 安装配置

HAProxy 安装配置步骤相对比较简单，跟其他源码软件安装方法大致相同，以下为 HAProxy 配置方法及步骤。

（1）HAProxy 编译及安装，操作指令如下，如图 11-2 所示。

```
cd /usr/src
wget -c https://www.haproxy.org/download/2.1/src/haproxy-2.1.12.tar.gz
yum install wget gcc -y
tar xzf haproxy-2.1.12.tar.gz
cd haproxy-2.1.12
make  TARGET=linux310  PREFIX=/usr/local/haproxy/
make  install  PREFIX=/usr/local/haproxy
```

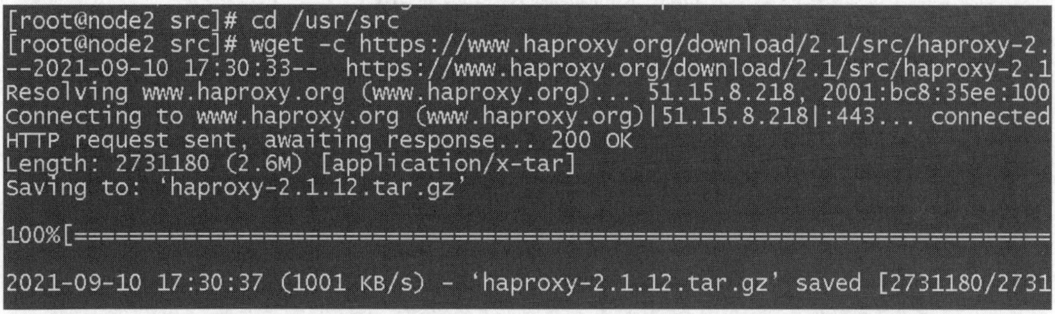

(a)

(b)

图 11-2　Kubernetes HAProxy 部署实战 1

（2）配置 HAProxy 服务，操作指令如下，如图 11-3 所示。

```
useradd -s /sbin/nologin haproxy -M
cd /usr/local/haproxy ;mkdir -p etc/
touch /usr/local/haproxy/etc/haproxy.cfg
cd /usr/local/haproxy/etc/
```

```
'doc/SPOE.txt' -> '/usr/local/haproxy/doc/haproxy/SPOE.txt'
'doc/intro.txt' -> '/usr/local/haproxy/doc/haproxy/intro.txt'
[root@master1 haproxy-2.1.12]#
[root@master1 haproxy-2.1.12]#
[root@master1 haproxy-2.1.12]#
[root@master1 haproxy-2.1.12]# useradd -s /sbin/nologin haproxy -M
cd /usr/local/haproxy ;mkdir -p etc/
touch /usr/local/haproxy/etc/haproxy.cfg
cd /usr/local/haproxy/etc/[root@master1 haproxy-2.1.12]# cd /usr/local/h
[root@master1 haproxy]# touch /usr/local/haproxy/etc/haproxy.cfg
[root@master1 haproxy]# cd /usr/local/haproxy/etc/
[root@master1 etc]#
[root@master1 etc]#
```

图 11-3　Kubernetes HAProxy 部署实战 2

（3）haproxy.cfg 配置文件内容如下，如图 11-4 所示。

```
global
    log /dev/log    local0
    log /dev/log    local1 notice
    chroot /usr/local/haproxy
    stats socket /usr/local/haproxy/haproxy-admin.sock mode 660 level admin
    stats timeout 30s
    user haproxy
    group haproxy
    daemon
    nbproc 1
defaults
    log     global
    timeout connect 5000
    timeout client  10m
    timeout server  10m
listen admin_stats
    bind 0.0.0.0:10080
    mode http
    log 127.0.0.1 local0 err
    stats refresh 30s
    stats uri /admin
    stats realm welcome login\ Haproxy
    stats auth admin:123456
    stats hide-version
    stats admin if TRUE
listen kube-master
    bind 0.0.0.0:8443
    mode tcp
    option tcplog
```

```
    balance source
        server master1 192.168.1.145:6443 check inter 2000 fall 2 rise 2 weight 1
        server master2 192.168.1.146:6443 check inter 2000 fall 2 rise 2 weight 1
        server master3 192.168.1.147:6443 check inter 2000 fall 2 rise 2 weight 1
```

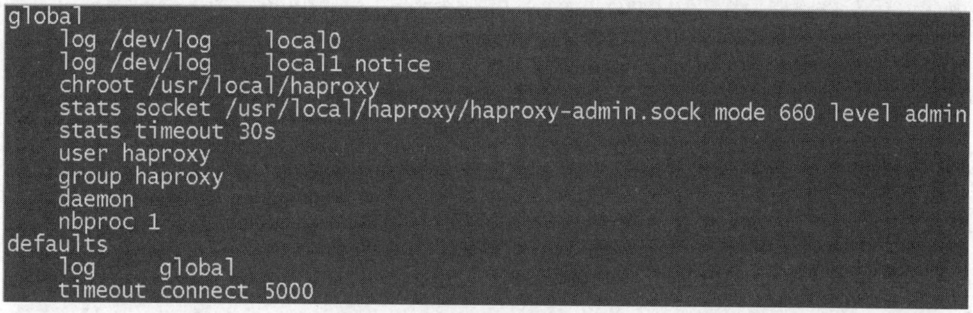

```
global
    log /dev/log     local0
    log /dev/log     local1 notice
    chroot /usr/local/haproxy
    stats socket /usr/local/haproxy/haproxy-admin.sock mode 660 level admin
    stats timeout 30s
    user haproxy
    group haproxy
    daemon
    nbproc 1
defaults
    log     global
    timeout connect 5000
```

(a)

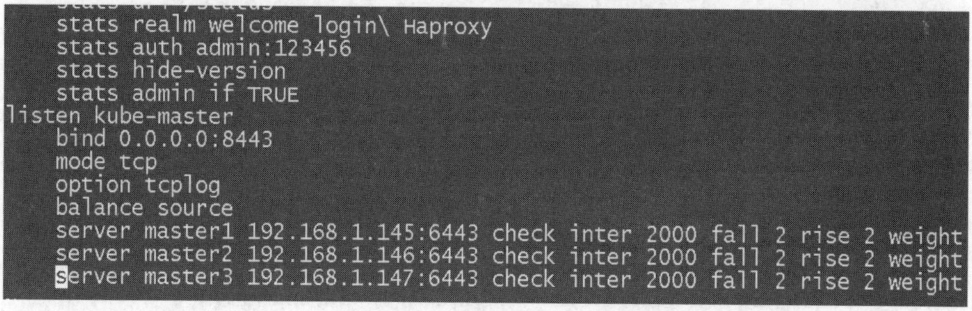

```
    stats uri /stats
    stats realm welcome login\ Haproxy
    stats auth admin:123456
    stats hide-version
    stats admin if TRUE
listen kube-master
    bind 0.0.0.0:8443
    mode tcp
    option tcplog
    balance source
    server master1 192.168.1.145:6443 check inter 2000 fall 2 rise 2 weight
    server master2 192.168.1.146:6443 check inter 2000 fall 2 rise 2 weight
    server master3 192.168.1.147:6443 check inter 2000 fall 2 rise 2 weight
```

(b)

图 11-4　Kubernetes HAProxy 部署实战 3

（4）启动 HAProxy 服务，操作指令如下，如图 11-5 所示。

```
/usr/local/haproxy/sbin/haproxy   -f   /usr/local/haproxy/etc/haproxy.cfg
ps -ef|grep -aiE haproxy
netstat -tnlp|grep -aiE haproxy
```

```
"haproxy.cfg" 33L, 925C written
[root@master1 etc]# /usr/local/haproxy/sbin/haproxy   -f   /usr/local/haprox
[root@master1 etc]# ps -ef|grep -aiE haproxy
haproxy   50632     1  0 17:34 ?        00:00:00 /usr/local/haproxy/sbin/hap
root      50714 29572  0 17:34 pts/0    00:00:00 grep --color=auto -aiE hapr
[root@master1 etc]#
[root@master1 etc]# netstat -tnlp|grep -aiE haproxy
tcp        0      0 0.0.0.0:8443            0.0.0.0:*               LISTEN
tcp        0      0 0.0.0.0:10080           0.0.0.0:*               LISTEN
[root@master1 etc]#
[root@master1 etc]#
[root@master1 etc]#
```

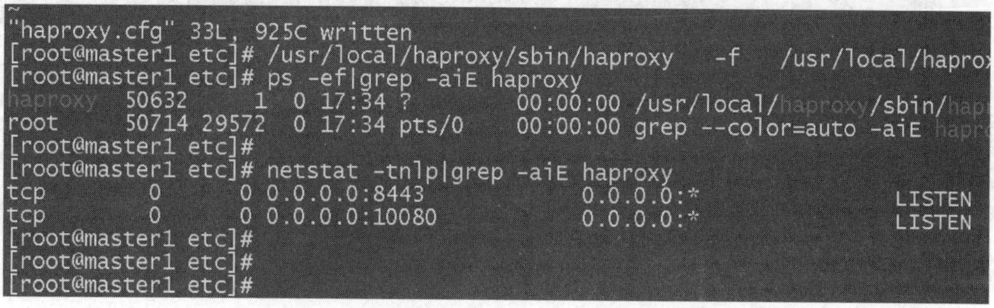

图 11-5　Kubernetes HAProxy 部署实战 4

（5）启动 HAProxy，报错如下：

```
[WARNING] 217/202150 (2857) : Proxy 'chinaapp.sinaapp.com': in multi-process mode, stats will be limited to process assigned to the current request.
```

解决方法：修改源码配置 src/cfgparse.c，找到以下行，调整 nbproc > 1 数值即可。

```
if (nbproc > 1) {
                if (curproxy->uri_auth) {
-                       Warning("Proxy '%s': in multi-process mode, stats will be limited to process assigned to the current request.\n",
+                       Warning("Proxy '%s': in multi-process mode, stats will be limited to the process assigned to the current request.\n",
```

11.4 配置 Keepalived 服务

Keepalived 是一个类似于工作在 layer 3、layer 4 和 layer 7 交换机制的软件，该软件有两种功能，分别是健康检查、VRRP 冗余协议。Keepalived 是模块化设计，不同模块负责不同的功能。部署 Keepalived 命令如下，如图 11-6 所示。

```
cd /usr/src;
yum install openssl-devel popt* kernel kernel-devel -y
wget -c http://www.keepalived.org/software/keepalived-1.2.1.tar.gz
tar xzf keepalived-1.2.1.tar.gz
cd keepalived-1.2.1 &&
./configure&&make &&make install
DIR=/usr/local/;\cp $DIR/etc/rc.d/init.d/keepalived /etc/rc.d/init.d/
\cp $DIR/etc/sysconfig/keepalived /etc/sysconfig/
mkdir -p /etc/keepalived && \cp $DIR/sbin/keepalived /usr/sbin/
cd /etc/keepalived/;touch keepalived.conf
```

```
make[1]: Leaving directory `/usr/src/keepalived-1.2.1/keepalived'
make -C genhash install
make[1]: Entering directory `/usr/src/keepalived-1.2.1/genhash'
install -d /usr/local/bin
install -m 755 ../bin/genhash /usr/local/bin/
install -d /usr/local/share/man/man1
install -m 644 ../doc/man/man1/genhash.1 /usr/local/share/man/man1
make[1]: Leaving directory `/usr/src/keepalived-1.2.1/genhash'
[root@node2 keepalived-1.2.1]# DIR=/usr/local/ ;\cp $DIR/etc/rc.d/in
[root@node2 keepalived-1.2.1]# \cp $DIR/etc/sysconfig/keepalived /et
[root@node2 keepalived-1.2.1]# mkdir -p /etc/keepalived && \cp $DIR
[root@node2 keepalived-1.2.1]#
```

图 11-6 Keepalived 案例实战

11.5　Keepalived master 配置实战

HAProxy+Keepalived master 端 keepalived.conf 配置文件如下：

```
! Configuration File for keepalived
global_defs {
notification_email {
    wgkgood@139.com
}
  notification_email_from wgkgood@139.com
  smtp_server 127.0.0.1
  smtp_connect_timeout 30
  router_id LVS_DEVEL
}
vrrp_script chk_haproxy {
  script "/data/sh/check_haproxy.sh"
  interval 2
  weight 2
}
# VIP1
vrrp_instance VI_1 {
    state MASTER
    interface ens33
    virtual_router_id 151
    priority 100
    advert_int 5
    nopreempt
    authentication {
       auth_type  PASS
       auth_pass  2222
    }
    virtual_ipaddress {
       192.168.1.188
    }
    track_script {
    chk_haproxy
    }
}
```

11.6 Keepalived Backup 配置实战

（1）HAProxy+Keepalived Backup 端 master2，配置 keepalived.conf，同时修改优先级为 90，操作指令如下：

```
! Configuration File for keepalived
global_defs {
notification_email {
    wgkgood@139.com
}
  notification_email_from wgkgood@139.com
  smtp_server 127.0.0.1
  smtp_connect_timeout 30
  router_id LVS_DEVEL
}
vrrp_script chk_haproxy {
  script "/data/sh/check_haproxy.sh"
  interval 2
  weight 2
}
# VIP1
vrrp_instance VI_1 {
    state BACKUP
    interface ens33
    virtual_router_id 151
    priority 90
    advert_int 5
    nopreempt
    authentication {
        auth_type PASS
        auth_pass 2222
    }
    virtual_ipaddress {
        192.168.1.188
    }
    track_script {
    chk_haproxy
    }
}
```

（2）HAProxy+Keepalived Backup 端 master3，配置 keepalived.conf，同时修改优先级为 80，操作指令如下：

```
! Configuration File for keepalived
global_defs {
notification_email {
    wgkgood@139.com
}
  notification_email_from wgkgood@139.com
  smtp_server 127.0.0.1
  smtp_connect_timeout 30
  router_id LVS_DEVEL
}
vrrp_script chk_haproxy {
  script "/data/sh/check_haproxy.sh"
  interval 2
  weight 2
}
# VIP1
vrrp_instance VI_1 {
    state  BACKUP
    interface ens33
    virtual_router_id 151
    priority 80
    advert_int 5
    nopreempt
    authentication {
        auth_type  PASS
        auth_pass  2222
    }
    virtual_ipaddress {
        192.168.1.188
    }
    track_script {
    chk_haproxy
    }
}
```

11.7 创建 HAProxy 检查脚本

在/data/sh/目录下创建 HAProxy 服务检查脚本，脚本名称为 check_haproxy.sh，同时设置执行权限。脚本内容如下：

```
mkdir -p /data/sh/
cd /data/sh/
touch check_haproxy.sh
chmod +x check_haproxy.sh
cat>/data/sh/check_haproxy.sh<<EOF
#!/bin/bash
#auto check haprox process
#2021-11-12  jfedu.net
CHECK_NUM=\$(ps -ef|grep -aiE haproxy|grep -aicvE "check|grep")
if
   [[ \$CHECK_NUM -eq 0 ]];then
   /etc/init.d/keepalived stop
fi
EOF
```

11.8 HAProxy+Keepalived 验证

启动所有节点 Keepalived 服务，然后查看 145 的 Keepalived 后台日志，且访问 188 VIP 正常，确认其可提供服务，即证明 HAProxy+Keepalived 高可用架构配置完毕，如图 11-7 所示。

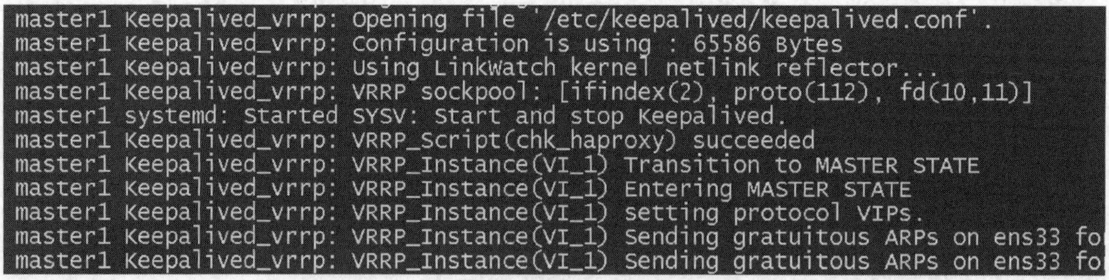

```
master1 Keepalived_vrrp: Opening file '/etc/keepalived/keepalived.conf'.
master1 Keepalived_vrrp: Configuration is using : 65586 Bytes
master1 Keepalived_vrrp: Using LinkWatch kernel netlink reflector...
master1 Keepalived_vrrp: VRRP sockpool: [ifindex(2), proto(112), fd(10,11)]
master1 systemd: Started SYSV: Start and stop Keepalived.
master1 Keepalived_vrrp: VRRP_Script(chk_haproxy) succeeded
master1 Keepalived_vrrp: VRRP_Instance(VI_1) Transition to MASTER STATE
master1 Keepalived_vrrp: VRRP_Instance(VI_1) Entering MASTER STATE
master1 Keepalived_vrrp: VRRP_Instance(VI_1) setting protocol VIPs.
master1 Keepalived_vrrp: VRRP_Instance(VI_1) Sending gratuitous ARPs on ens33 fo
master1 Keepalived_vrrp: VRRP_Instance(VI_1) Sending gratuitous ARPs on ens33 fo
```

(a)

图 11-7　HAProxy+Keepalived 网站架构

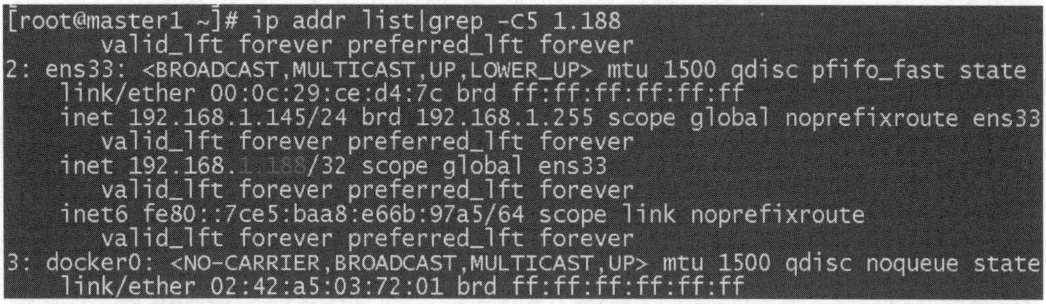

```
[root@master1 ~]# ip addr list|grep -c5 1.188
       valid_lft forever preferred_lft forever
2: ens33: <BROADCAST,MULTICAST,UP,LOWER_UP> mtu 1500 qdisc pfifo_fast state
    link/ether 00:0c:29:ce:d4:7c brd ff:ff:ff:ff:ff:ff
    inet 192.168.1.145/24 brd 192.168.1.255 scope global noprefixroute ens33
       valid_lft forever preferred_lft forever
    inet 192.168.1.188/32 scope global ens33
       valid_lft forever preferred_lft forever
    inet6 fe80::7ce5:baa8:e66b:97a5/64 scope link noprefixroute
       valid_lft forever preferred_lft forever
3: docker0: <NO-CARRIER,BROADCAST,MULTICAST,UP> mtu 1500 qdisc noqueue state
    link/ether 02:42:a5:03:72:01 brd ff:ff:ff:ff:ff:ff
```

(b)

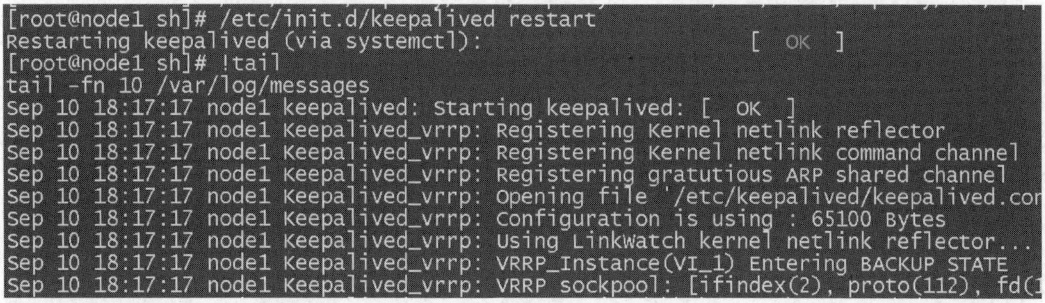

```
[root@node1 sh]# /etc/init.d/keepalived restart
Restarting keepalived (via systemctl):           [  OK  ]
[root@node1 sh]# !tail
tail -fn 10 /var/log/messages
Sep 10 18:17:17 node1 keepalived: Starting keepalived: [ OK ]
Sep 10 18:17:17 node1 Keepalived_vrrp: Registering Kernel netlink reflector
Sep 10 18:17:17 node1 Keepalived_vrrp: Registering Kernel netlink command channel
Sep 10 18:17:17 node1 Keepalived_vrrp: Registering gratutious ARP shared channel
Sep 10 18:17:17 node1 Keepalived_vrrp: Opening file '/etc/keepalived/keepalived.con
Sep 10 18:17:17 node1 Keepalived_vrrp: Configuration is using : 65100 Bytes
Sep 10 18:17:17 node1 Keepalived_vrrp: Using LinkWatch kernel netlink reflector...
Sep 10 18:17:17 node1 Keepalived_vrrp: VRRP_Instance(VI_1) Entering BACKUP STATE
Sep 10 18:17:17 node1 Keepalived_vrrp: VRRP sockpool: [ifindex(2), proto(112), fd(1
```

(c)

```
[root@node1 ~]# !tail
tail -fn 10 /var/log/messages
Sep 10 18:17:17 node1 Keepalived_vrrp: Configuration is using : 65100 Bytes
Sep 10 18:17:17 node1 Keepalived_vrrp: Using LinkWatch kernel netlink reflector..
Sep 10 18:17:17 node1 Keepalived_vrrp: VRRP_Instance(VI_1) Entering BACKUP STATE
Sep 10 18:17:17 node1 Keepalived_vrrp: VRRP sockpool: [ifindex(2), proto(112), fd
Sep 10 18:17:17 node1 Keepalived_vrrp: VRRP_Script(chk_haproxy) succeeded
Sep 10 18:24:51 node1 Keepalived_vrrp: VRRP_Instance(VI_1) Transition to MASTER S
Sep 10 18:24:56 node1 Keepalived_vrrp: VRRP_Instance(VI_1) Entering MASTER STATE
Sep 10 18:24:56 node1 Keepalived_vrrp: VRRP_Instance(VI_1) setting protocol VIPs.
Sep 10 18:24:56 node1 Keepalived_vrrp: VRRP_Instance(VI_1) Sending gratuitous ARP
Sep 10 18:25:01 node1 Keepalived_vrrp: VRRP_Instance(VI_1) Sending gratuitous ARP
```

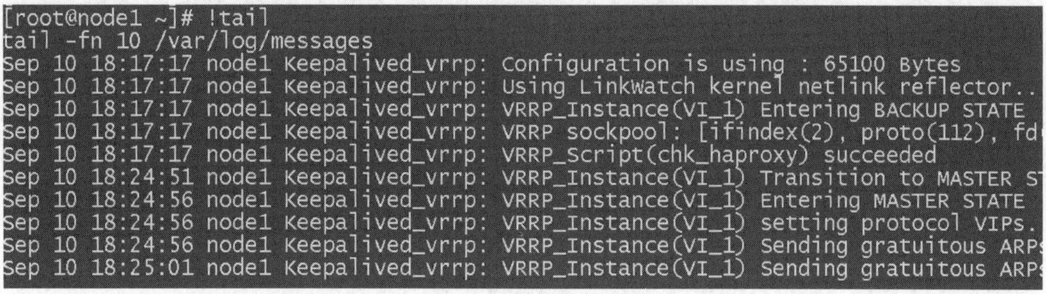

(d)

图 11-7 （续）

11.9 初始化 master 集群

（1）Kubernetes 集群引入 HAProxy 高可用集群，可以生成 Kubernetes 集群 init 配置文件，操作指令如下：

```
kubeadm config print init-defaults >kubeadmin-init.yaml
```

（2）根据以上指令生成配置文件之后，修改 kubeadmin-init.yaml 配置文件，添加集群

controlPlaneEndpoint: "192.168.1.188:8443"，最终代码如下：

```
apiVersion: kubeadm.Kubernetes.io/v1beta2
bootstrapTokens:
- groups:
  - system:bootstrappers:kubeadm:default-node-token
  token: abcdef.0123456789abcdef
  ttl: 24h0m0s
  usages:
  - signing
  - authentication
kind: InitConfiguration
localAPIEndpoint:
  advertiseAddress: 192.168.1.145
  bindPort: 6443
nodeRegistration:
  criSocket: /var/run/dockershim.sock
  name: master1
  taints:
  - effect: NoSchedule
    key: node-role.kubernetes.io/master
---
apiServer:
  timeoutForControlPlane: 4m0s
apiVersion: kubeadm.Kubernetes.io/v1beta2
certificatesDir: /etc/kubernetes/pki
clusterName: kubernetes
controlPlaneEndpoint: "192.168.1.188:8443"
controllerManager: {}
dns:
  type: CoreDNS
etcd:
  local:
    dataDir: /var/lib/etcd
imageRepository: registry.aliyuncs.com/google_containers
kind: ClusterConfiguration
kubernetesVersion: v1.20.0
networking:
  dnsDomain: cluster.local
  podSubnet: 10.244.0.0/16
  serviceSubnet: 10.10.0.0/16
scheduler: {}
```

（3）执行以下命令初始化集群即可，如图 11-8 所示，操作指令如下：

```
kubeadm init --config kubeadmin-init.yaml --upload-certs
```

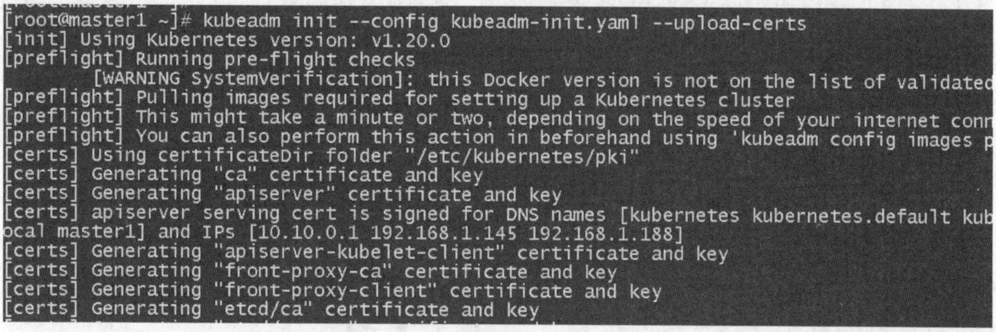

```
[root@master1 ~]# kubeadm init --config kubeadm-init.yaml --upload-certs
[init] Using Kubernetes version: v1.20.0
[preflight] Running pre-flight checks
        [WARNING SystemVerification]: this Docker version is not on the list of validated
[preflight] Pulling images required for setting up a Kubernetes cluster
[preflight] This might take a minute or two, depending on the speed of your internet conn
[preflight] You can also perform this action in beforehand using 'kubeadm config images p
[certs] Using certificateDir folder "/etc/kubernetes/pki"
[certs] Generating "ca" certificate and key
[certs] Generating "apiserver" certificate and key
[certs] apiserver serving cert is signed for DNS names [kubernetes kubernetes.default kub
ocal master1] and IPs [10.10.0.1 192.168.1.145 192.168.1.188]
[certs] Generating "apiserver-kubelet-client" certificate and key
[certs] Generating "front-proxy-ca" certificate and key
[certs] Generating "front-proxy-client" certificate and key
[certs] Generating "etcd/ca" certificate and key
```

(a)

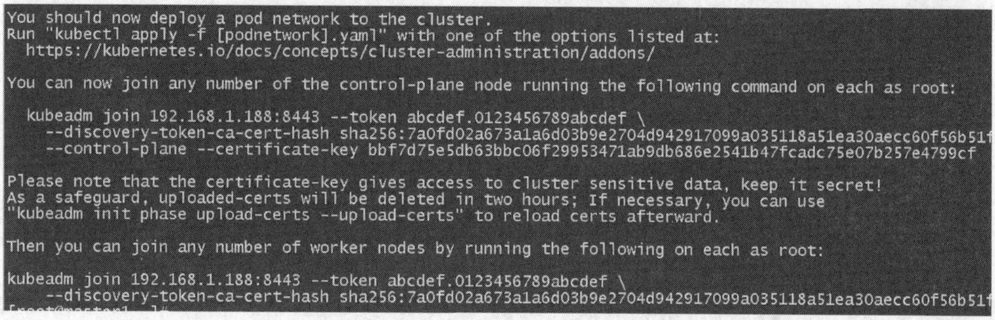

```
You should now deploy a pod network to the cluster.
Run "kubectl apply -f [podnetwork].yaml" with one of the options listed at:
  https://kubernetes.io/docs/concepts/cluster-administration/addons/

You can now join any number of the control-plane node running the following command on each as root:

  kubeadm join 192.168.1.188:8443 --token abcdef.0123456789abcdef \
    --discovery-token-ca-cert-hash sha256:7a0fd02a673a1a6d03b9e2704d942917099a035118a51ea30aecc60f56b51f \
    --control-plane --certificate-key bbf7d75e5db63bbc06f29953471ab9db686e2541b47fcadc75e07b257e4799cf

Please note that the certificate-key gives access to cluster sensitive data, keep it secret!
As a safeguard, uploaded-certs will be deleted in two hours; If necessary, you can use
"kubeadm init phase upload-certs --upload-certs" to reload certs afterward.

Then you can join any number of worker nodes by running the following on each as root:

kubeadm join 192.168.1.188:8443 --token abcdef.0123456789abcdef \
    --discovery-token-ca-cert-hash sha256:7a0fd02a673a1a6d03b9e2704d942917099a035118a51ea30aecc60f56b51f
[root@master1 ~]#
```

(b)

图 11-8　Kubernetes 多 master 集群初始化

（4）如果集群只有 3 台 master，没有 node 节点，可以去除 master 节点上的污点标记，使其可以分配 Pod 资源，操作指令如下：

```
kubectl taint nodes --all node-role.kubernetes.io/master-
```

11.10　Kubernetes Dashboard UI 实战

Kubernetes 实现的最重要的工作是对 Docker 容器集群进行统一管理和调度，通常使用命令行操作 Kubernetes 集群及各个节点。命令行操作非常不方便，如果使用 UI 界面进行可视化操作，会更加方便管理和维护。以下为配置 Kubernetes Dashboard 的完整过程：

（1）下载 Dashboard 配置文件。

```
wget https://raw.githubusercontent.com/kubernetes/dashboard/v2.0.0-rc5/
```

```
aio/deploy/recommended.yaml
\cp recommended.yaml recommended.yaml.bak
```

（2）修改文件 recommended.yaml 的 39 行内容。因为默认情况下，service 的类型是 cluster IP，需更改为 NodePort 的方式，便于访问，也可映射到指定的端口。

```
spec:
  type: NodePort
  ports:
    - port: 443
      targetPort: 8443
      nodePort: 31001
  selector:
    Kubernetes-app: kubernetes-dashboard
```

（3）修改文件 recommended.yaml 的 195 行内容，因为默认情况下 Dashboard 为英文显示，可以设置为中文。

```
env:
         - name: ACCEPT_LANGUAGE
           value: zh
```

（4）创建 Dashboard 服务，指令操作如下：

```
kubectl apply -f recommended.yaml
```

（5）查看 Dashboard 运行状态。

```
kubectl get pod -n kubernetes-dashboard
kubectl get svc -n kubernetes-dashboard
```

（6）将 master 节点也设置为 node 节点，可以运行 Pod 容器任务，命令如下：

```
kubectl taint nodes --all node-role.kubernetes.io/master-
```

（7）基于 Token 的方式访问，设置和绑定 Dashboard 权限，命令如下：

```
#创建 Dashboard 的管理用户
kubectl create serviceaccount dashboard-admin -n kube-system
#将创建的 Dashboard 用户绑定为管理用户
kubectl create clusterrolebinding dashboard-cluster-admin --clusterrole=cluster-admin --serviceaccount=kube-system:dashboard-admin
#获取刚刚创建的用户对应的 Token 名称
kubectl get secrets -n kube-system | grep dashboard
#查看 Token 的详细信息
kubectl describe secrets -n kube-system $(kubectl get secrets -n kube-system | grep dashboard |awk '{print $1}')
```

（8）通过浏览器访问 Dashboard Web，https://192.168.1.188:31001/，输入 Token 登录即可，如图 11-9 所示。

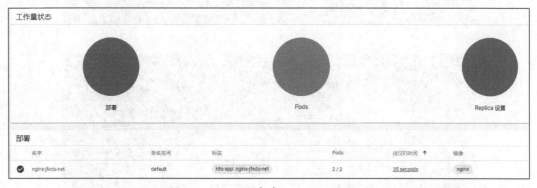

图 11-9　Kubernetes Web 界面展示

第 12 章 Kubernetes 配置故障实战

12.1 etcd 配置中心故障错误一

配置完毕却无法启动，如图 12-1 所示。

```
##########
# [member]
ETCD_NAME=etcd3
ETCD_DATA_DIR="/data/etcd"
#ETCD_WAL_DIR=""
#ETCD_SNAPSHOT_COUNT="10000"
#ETCD_HEARTBEAT_INTERVAL="100"
#ETCD_ELECTION_TIMEOUT="1000"
#ETCD_LISTEN_PEER_URLS="http://192.168.0.133:2380"
ETCD_LISTEN_CLIENT_URLS="http://192.168.0.133:2379,http://127.0.0.1:2379"
CD_MAX_SNAPSHOTS="5"
#ETCD_MAX_WALS="5"
#ETCD_CORS=""
#[cluster]
ETCD_INITIAL_ADVERTISE_PEER_URLS="http://192.168.0.133:2380"
# if you use different ETCD_NAME (e.g. test), set ETCD_INITIAL_CLUSTER value for this nam
ETCD_INITIAL_CLUSTER="etcd1=http://192.168.0.131:2380,etcd2=http://192.168.0.132:2380,etc
#ETCD_INITIAL_CLUSTER_STATE="new"
#ETCD_INITIAL_CLUSTER_TOKEN="etcd-cluster"
ETCD_ADVERTISE_CLIENT_URLS="http://192.168.0.133:2379"
#ETCD_DISCOVERY=""
#ETCD_DISCOVERY_SRV=""
```

（a）

```
 etcd[40945]: setting maximum number of CPUs to 1, total number of available CPUs
 etcd[40945]: listening for peers on http://192.168.0.133:2380
 etcd[40945]: listening for client requests on 127.0.0.1:2379
 etcd[40945]: listening for client requests on 192.168.0.133:2379
 etcd[40945]: cannot access data directory: mkdir /data/etcd: permission denied
 systemd[1]: etcd.service: main process exited, code=exited, status=1/FAILURE
 systemd[1]: Failed to start Etcd Server.
s failed

ktop.org/mailman/listinfo/systemd-devel

 systemd[1]: Unit etcd.service entered failed state.
 systemd[1]: etcd.service failed.
 systemd[1]: etcd.service holdoff time over, scheduling restart.
 systemd[1]: start request repeated too quickly for etcd.service
 systemd[1]: Failed to start Etcd Server.
```

（b）

图 12-1　etcd 启动服务报错

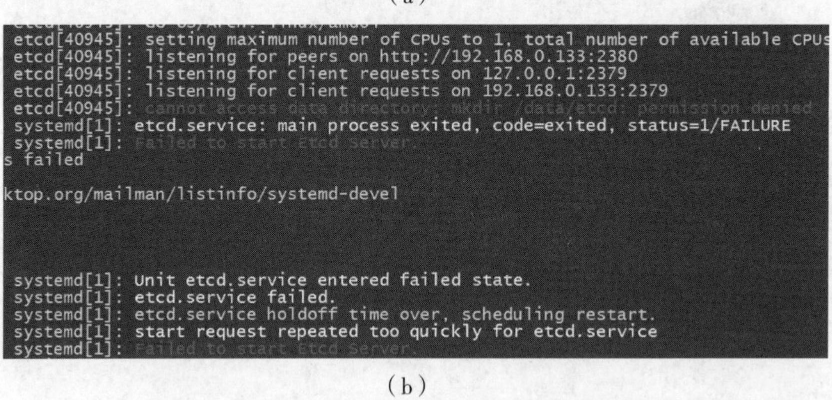

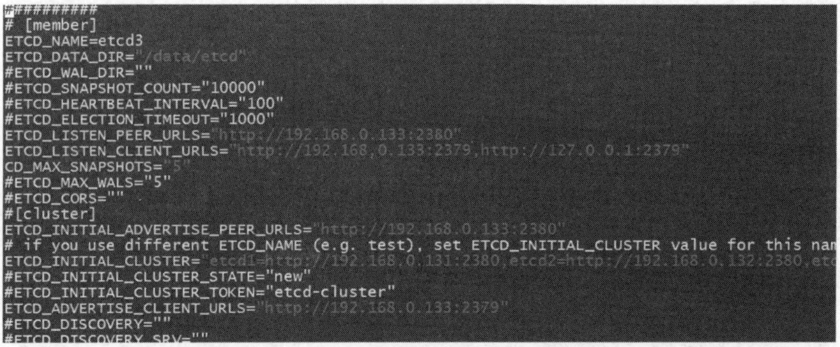

（1）原因分析：根据日志提示，启动 etcd 服务时，会初始化 etcd 数据目录/data/etcd/，而该目录不存在，自动创建目录提示权限被拒绝。

（2）解决方法如下：

① 创建/data/etcd/配置目录，命令为 mkdir –p /data/etcd/。

② 同时设置权限为可写，命令为 chmod 757 –R /data/etcd/。

12.2　etcd 配置中心故障错误二

部署 etcd 集群时，通常尝试启动 etcd1 时，报错信息如下：

```
Apr 12 01:06:49 Kubernetes-master etcd[3092]: health check for peer 618d69366dd8cee3 could not connect: dial tcp 10.0.0.122:2380: getsockopt: connection refused
Apr 12 01:06:49 Kubernetes-master etcd[3092]: health check for peer acd2ba924953b1ec could not connect: dial tcp 192.168.0.143: 2380: getsockopt: connection refused
Apr 12 01:06:48 Kubernetes -master etcd[3092]: publish error: etcdserver: request timed out
```

（1）原因分析：etcd1 的配置文件/etc/etcd/etcd.conf 中的 etcd_INITIAL_CLUSTER_STATE 是 new，而在配置中 etcd_INITIAL_CLUSTER 写入了 etcd2、etcd3 的 IP:PORT，这时 etcd1 尝试连接 etcd2、etcd3，但是 etcd2、etcd3 的 etcd 服务此时还未启动。

（2）解决方法：启动 etcd2 和 etcd3 的 etcd 服务，再启动 etcd1。

12.3　Pod infrastructure 故障错误三

Kubernetes Pod–infrastructure:latest 镜像下载失败，报错如下：

```
image pull failed for registry.access.redhat.com/rhel7/pod-infrastructure: latest, this may be because there are no credentials on this request.  details: (open /etc/docker/certs.d/registry.access.redhat.com/redhat-ca.crt: no such file or directory)
Failed to create pod infra container: ImagePullBackOff; Skipping pod "redis-master-jj6jw_default(fec25a87-cdbe-11e7-ba32-525400cae48b)": Back-off pulling image "registry.access.redhat.com/rhel7/pod-infrastructure: latest
```

报错原因：因为获取远程 Pod 镜像需要查找安全证书，所以需要从官网下载证书，但由于网络故障无法获取远程镜像。解决方法如下：

```
yum install *rhsm* -y
vim /etc/kubernetes/kubelet,pod 外网源如下：
KUBELET_POD_INFRA_CONTAINER="--pod-infra-container-image=registry.access
.redhat.com/rhel7/pod-infrastructure:latest"
可以更改为内网的私有源即可；
KUBELET_POD_INFRA_CONTAINER="--pod-infra-container-image=192.168.1.146:
5000/rhel7/pod-infrastructure:latest"
```

12.4　Docker 虚拟化故障错误四

在 Kubernetes 节点上维护管理 Docker 时，node 节点的 Docker 无法启动，报错信息如下：

```
systemctl start docker.service
Job for docker.service failed because the control process exited with error
code. See "systemctl status docker.service" and "journalctl -xe" for details.
```

（1）错误原因：因为找不到 devicemap 存储块相关的库，导致无法启动 Docker，可以通过安装 devcie-map 相关的库解决问题。

（2）解决方法如下：

```
yum install device-map* -y
```

12.5　Docker 虚拟化故障错误五

Kubernetes 客户端 node 节点，部署 Docker 服务，发现启动 Docker 容器时报错如下：

```
/usr/bin/docker-current: Error response from daemon: shim error: docker-runc
not installed on system.
```

（1）原因分析：可能是 Docker 的版本 bug，Docker 软件安装不正确，导致无法启动。

（2）解决方法：更换 yum 源，卸载当前 Docker，重新安装启动，Docker 容器可以正常运行。

12.6　Dashboard API 故障错误六

当配置好 Dashborad Web UI 时，浏览器访问 UI：http://10.0.0.122:8080/ui，显示的是相关的提示信息，而不是 Kebernetes Dashboard 的页面。

（1）原因分析：检查一下 master 端，重点检查 API server 接口问题。

（2）解决方法：在 master 端修改 /etc/kubernetes/apiserver 中的配置文件，删除 ServiceAccount，

重新访问 Dashboard 界面如下：

```
KUBE_ADMISSION_CONTROL="--admission-control=NamespaceLifecycle,Namespace
Exists,LimitRanger,SecurityContextDeny,ResourceQuota"
```

12.7 Dashboard 网络访问故障错误七

当配置好 Dashborad Web UI 后，浏览器访问 UI：http://10.0.0.122:8080/ui，显示的是相关的提示信息，而不是 Kebernetes Dashboard 的页面，修改 API server 也无法解决该问题，最终访问 UI 超时，显示如下信息：

```
Error: 'dial tcp 10.0.66.2:9090: getsockopt: connection timed out'
Trying to reach: 'http://10.0.66.2:9090/'
```

（1）原因分析：远程服务器 10.0.66.2 的 9090 端口无法连接，应该属于网络的问题。

（2）解决方法：重启 Kubernetes 集群相关服务即可，命令操作如下：

```
service etcd restart;service flanneld restart;service docker restart;
iptables -P FORWARD ACCEPT
```

通过 ifconfig 查看各个服务器的 IP 信息，同时确认 Flanneld 配置是否正确，最终解决问题，如图 12-2 所示。

图 12-2　Kubernetes Web 界面